Coal Age Scene in miniature exhibit in Museumobile

Forests of the coal age were densely populated with tree-ferns, peculiar seed-ferns, and gigantic lycopod trees. In the foreground may be seen an amphibious creature which inhabited the ancient swamplands.

LEAVES AND STEMS FROM FOSSIL FORESTS

A Handbook of the Paleobotanical Collections in the Illinois State Museum

By

RAYMOND E. JANSSEN

Illinois State Museum
Popular Science Series, Vol. I

Illinois State Museum
Springfield, Illinois
1979

First Printing, 1939
Second Printing (revised), 1957
Third Printing, 1965
Fourth Printing (revised), 1979

ISSN 0360-0297
ISBN 0-89792-077-5

Printed by authority of the State of Illinois

(P.O. 12585–2500–6/79)

2

POPULAR SCIENCE SERIES

Vol. I — Leaves and Stems from Fossil Forests.
Raymond E. Janssen. 1939. 4th Printing (revised), 1979.

Vol. II — Birds in Your Backyard.
Virginia S. Eifert. 1941. 2nd edition, revised by Milton D. Thompson, 1967.

Vol. III — Illinois Wild Flowers.
John Voss and Virginia S. Eifert. 1951. 2nd edition, 1978.

Vol. IV — A Distributional Check List of the Birds of Illinois.
Harry R. Smith and Paul W. Parmalee. 1955. Out of print. (Superseded by Volume IX.)

Vol. V — Reptiles of Illinois.
Paul W. Parmalee. 1955.

Vol. VI — Aquatic Plants of Illinois.
Glen S. Winterringer and Alvin C. Lopinot. 1966. 2nd printing (revised), 1977.

Vol. VII — Wild Orchids of Illinois.
Glen S. Winterringer. 1967.

Vol. VIII — The Fresh-water Mussels of Illinois.
Paul W. Parmalee. 1967.

Vol. IX — An Annotated Check-list of the Birds of Illinois.
H. David Bohlen. 1978.

LEAVES AND STEMS FROM FOSSIL FORESTS

A Handbook of the Paleobotanical Collections in the Illinois State Museum

By

Raymond E. Janssen

PREFACE

THE immediate stimulus that produced this book was derived from the Langford collection of fossils, probably the finest ever assembled from the Mazon Creek region of Illinois. The vast amount of labor entailed in gathering, selecting, opening and "developing" the specimens can be sensed in a measure from an excerpt of a letter from Mr. George Langford, Sr.

"George, Jr., and I spent 140 days (morning and afternoon) in the field . . . and much more time in washing, mending, sorting, marking and developing the material. Altogether I devoted . . . 3400 hours to all this, and George, Jr., 2000 . . ." Nearly 250,000 nodules were split open, only ten percent of which were taken away from the strip mine heaps. Of these 25,000, four hundred fifty-nine (of which sixteen are animals) were selected for the museum.

The excellent quality of the collection, both individually and as a unit, bespeaks scientific acumen, courage, and integrity, a special aptitude, and a tirelessness of purpose, that can be fully appreciated only by those who have engaged in similar labors. The Illinois State Museum is deeply indebted to the Langfords for the opportunity to exhibit, study and publish on this unique collection. To Mr. George Langford, Sr., now paleobotanist of the Chicago Natural History Museum, additional thanks and sincere appreciation are extended for his courtesy and cooperation in furnishing new negatives, photographic prints and valuable suggestions in the revision of illustrations in the new edition.

The author, Raymond E. Janssen, formerly paleobotanist, Illinois State Museum, now Professor and Head of the Department of Geology, Marshall College, Huntington, West Virginia, has been devoted twenty-five years to the study of the origin of coal and to carboniferous paleobotany. The excellence of the original text prepared twenty years ago, which required no important change or revision, is evidence of the high standard characteristic of Professor Janssen's work. He has the fortunate faculty of embodying scientific information in simple direct language and presenting it in an entertaining manner.

THORNE DEUEL
Museum Director

May 1, 1957.

CONTENTS

INTRODUCTION

THE EXTENSIVE coal fields of Illinois, in addition to providing the coal itself, have furnished science with a great store of information regarding the plants which flourished here in past ages and were responsible for making the coal. The widespread practice of "strip mining" in localities where coal beds lie near the surface has stimulated extensive collecting of plant fossils, particularly by the non-scientist. The need of a non-technical handbook, describing the more widely distributed forms within the state, first became apparent nearly twenty years ago. The original edition of this publication, illustrating and describing more than one hundred fifty genera and species of the most common fossil plants from Illinois, was prepared to fill that need. Its usefulness was not confined to Illinois, however, because the same types of plant fossils are found in all other regions where coal beds of the same geological age occur. In general, nearly all of the coal fields in the eastern half of the United States and Canada are of this same age. In response to continued demands for this handbook, a new edition has become necessary.

The specimens illustrated herein, for the most part, are on display or in the collections at the Illinois State Museum. Included in the Museum collections are many original "type" specimens obtained during the first geological survey of the state nearly a century ago, as well as the Langford assemblage of fossil plants from Will County. The latter, collected more recently by the George Langfords, father and son, formerly of Joliet, Illinois, is one of the finest and most representative collections in existence from that portion of the state, widely known as the famous "Mazon Creek" region. The Langford Collection was obtained from the strip mines over a period of several years by frequent visits to the field. In addition, the Langfords spent countless painstaking hours in cleaning and preparing the specimens so as to permit the individual plant remains to stand out in marked contrast from the rock matrix. The Langfords have performed an outstanding service in depositing this excellent collection in the Illinois State Museum for scientific study.

The specimens described in this handbook are not intended to represent any particular horizon or locality in Illinois. However, because of their excellent preservation, most of the illustrations have been made from specimens in the Langford Collection from the strip mines near Wilmington. For the same reason, a few specimens from foreign localities are pictured, but in each case the species occurs in Illinois also. Unless otherwise indicated, specimens illustrated came from the Mazon Creek region and were collected by Mr. George Langford, Sr. and Mr. George Langford, Jr. The photographs, unless specifically noted otherwise, were made by Mr. Charles W. Hodge of Springfield. Additional photographs were furnished by the Illinois State Geological Survey, by George Langford, Sr., Franklin C. MacKnight, and the author. All drawings are by the author. Photographs of restored trees and other plants of the coal age are reproduced by courtesy of the Chicago Natural History Museum, Chicago. Helpful assistance and suggestions were rendered in the preparation of this handbook by the late Professor A. C. Noé, paleobotanist at the University of Chicago, shortly before his untimely death. It is hoped that this second edition of LEAVES AND STEMS FROM FOSSIL FORESTS will continue to serve a useful purpose not only in Illinois, but wherever plant fossils are found in the coal-bearing strata of the Pennsylvanian Period.

RAYMOND E. JANSSEN.

Marshall College,
Huntington, West Virginia.
January 1, 1957.

CHAPTER I

Impressions From the Past

"A knowledge of the past, however imperfect, adds to the attractiveness of the present."
—A. C. Seward

THE LONGEVITY OF THE SEQUOIA, the majesty of the palm-tree, the beauty of the orchid, or the fragrance of the rose are qualities which we recognize, appreciate, and accept as an integral part of the world in which we live. When we realize that the plants of today are merely the end-products of a long line of dissimilar forms which flourished in bygone ages, their existence becomes more significant to us. For example, the most of us think of coal simply as a dirty black material taken from the earth and used to heat our homes or furnish power for the operation of our factories and steam engines. But when we recall that coal is composed almost entirely of the metamorphosed remains of plants and trees that lived in an earlier geological period, its consideration, together with the nature of the organisms which formed it, stimulates a new interest in our thoughts.

The common occurrence of petrified wood in many regions of the world is well known. Fossilized stems and sections of tree trunks are frequently found, the structural details having been made permanent by the deposition of siliceous or calcareous material from water, drawn by capillarity into the dead but still uncollapsed tissues. Impressive examples of petrified trees may be seen in Arizona and Yellowstone National Park. Petrified Forest National Monument in Arizona includes a region in which the ground over an area of many square miles is strewn with petrified tree trunks, some reaching lengths of two hundred feet and diameters of ten feet. In Yellowstone Park a thick mass of volcanic ash, known as Amethyst Mountain, has buried not one but several successive prehistoric forests as they grew. The deposit has preserved the upright trunks which have become exposed to view as erosion wears down the mountain side. In Sutherlandshire, Scotland, it is possible to pick up great quantities of petrified wood on the sea beach where the waves are cutting into strata containing large numbers of petrified trees. Occurrences of petrified wood throughout the world are too numerous to mention. In most cases such as these, only the woody portions of the trees have become preserved, and we know relatively little of the leaves, seeds, and herbaceous plant life that existed with them. However, when we look toward the coal-producing regions of the earth, we are able to view a more nearly complete picture of the plant life of ancient times. Illinois is fortunate in having within its boundaries some of the finest fossil plant deposits that exist anywhere in the world.

The earth's surface is made up in part of accumulations of old sediments. Some are indistinguishable from the rubble, sand, and mud now in process of accumulation in the lowlands from the endless processes of erosion taking place day by day. Other accumulations of sediments have become hardened through the ages and may be gently folded or violently broken, contorted and uplifted by crustal movements so as to render their sedimentary origin almost unrecognizable. It is in these sediments of past ages—the sole remains of lost continents—that we find the buried fragments of the plants and animals which once constituted life on earth. It is from this record that we discover the antiquity of the orders of living plants or recognize extinct types which were connecting links between primitive and more modern forms. Because coal is a sedimentary rock having its origin in the plant life of another age, it follows that the study of coal should tell us much about the kinds of plants which formed it. However, the plant remains constituting the actual coal itself usually have been so squeezed and altered that their details are no longer recognizable. On the other hand, the remains which exist as fossils in the shale beds immediately above or below the coal seams are often so well preserved that complete leaves with their intricate veins, surface hairs and texture, and individual plant cells may be seen and studied.

The underlying beds of shale represent the former soil in which these plants grew. Commonly their roots are preserved just as they existed in the ancient soils. The shaley layers overlaying the coal beds represent the sediments which were deposited above the plants during times of flood, killing and burying them but preserving their remains as coal and fossils. Each layer of coal represents a former forest, and each layer of shale above it represents an interval when the sea advanced and flooded the swamplands. In Illinois sixteen different coal seams indicate that the sea rose and fell a corresponding number of times. Whenever the water-level dropped a few feet, the forests flourished; when it rose again, these forests were destroyed and buried. As later ages passed in slow succession, the enclosing sediments of mud and sand hardened into stone, and the buried plant material turned to coal.

The importance of coal in our civilization has caused the eyes of science to be directed toward its origin, with the result that much has been learned concerning the great forests of the past which have provided us with the coal fields of today. Coal has been deposited in various quantities ever since land plants began to flourish upon the earth. The most important of these coal deposits were formed during the Carboniferous period of the earth's history, or more specifically during the Pennsylvanian subdivision of this period. The Pennsylvanian period began about a quarter of a billion years ago and lasted for thirty-five million years. During this long period of time, great swamp forests of luxurious tropical vegetation covered much of the earth's surface, particularly in what is now the central and eastern portions of the United States, Europe,

THE GEOLOGIC TIME TABLE
(Read From Bottom Up)

ERA	PERIOD	BEGINNING DATE Years Ago	DOMINANT LIFE
Cenozoic	Recent	50,000	The age of man
	Pleistocene	1,000,000	Great Ice Age—man appears
	Tertiary	65,000,000	The age of mammals
Mesozoic	Cretaceous	135,000,000	Flowering plants appear
	Jurassic	190,000,000	Age of dinosaurs—birds appear
	Triassic	225,000,000	The age of reptiles
Paleozoic	Permian	280,000,000	The age of amphibians
	Pennsylvanian	325,000,000	The great coal-forming age
	Mississippian	345,000,000	The age of crinoids
	Devonian	400,000,000	The age of fishes
	Silurian	430,000,000	The age of corals
	Ordovician	500,000,000	The age of cephalopods
	Cambrian	570,000,000	The age of trilobites
Proterozoic Era		1,000,000,000	Age of algae and primitive invertebrates
Archeozoic Era		1,500,000,000	Age of larval life
Azoic Era		?	No known life

Earth history is divided by geologists into several major units—eras, and these are subdivided somewhat arbitrarily into periods. The time durations are estimates based on the rate of disintegration of radioactive minerals in the rocks. The evolution of life is based on fossil evidence found in the respective strata of the earth.

Coal Age Scene in miniature exhibit in Museumobile

FIGURE 1

A swamp forest of the coal age. Luxuriant tropical jungles such as this existed in Illinois a quarter of a billion years ago.

Asia, and even the northlands of Alaska, Greenland, and Siberia. The great coal fields of Illinois represent a small part of the vast accumulation of that age. The "Prairie State" was then a dense jungle of luxuriant vegetation!

Picture, if you can, an Illinois without its seas of waving cornfields, without its woodlands of deciduous broad-leaved trees, or its rippling grassy plains. In those days no brightly colored flowers added their brilliant hues to the landscapes; no little furry animals burrowed in the soil or scampered from tree to tree; no birds nested in the branches; and bees and butterflies were absent. Such forms of life had not yet made their appearance upon the earth, and were not destined to arrive until millions of years later. Instead, there existed vast swamplands in which abounded species of trees and plants which have long since become extinct. Although abundant, this vegetation consisted entirely of lower plant orders such as our present-day club mosses, horsetails, and ferns. Flowers, characteristic of higher orders, had not yet evolved. Although being lower forms botanically, many of these primitive plants were giants in size. The forests were dominated by great tree-like club mosses, huge horsetail rushes, and primitive gymnosperms which attained the dimensions of modern trees. The undergrowth consisted primarily of ferns and similar plants having fern-like foliage. From all indications, there were no seasonal variations in climate and growing conditions; hence the foliage was very dense, always green, and grew rapidly and abundantly. The only animals living in those forests were amphibians, small reptiles, and primitive kinds of insects. Huge dragonflies, some with wingspreads nearly thirty inches across, flew through the air, and giant cock-roaches crawled about over the logs and branches. In the waters lived fish, snails, clams, annelid worms, and crab-like crustaceans.

Our knowledge of these ancient plants is acquired from their fossil impressions found in the shale layers adjacent to the coal seams. Here the individual plant parts have been

FIGURE 2

An enormous electric shovel in operation at an Illinois strip coal mine. The great thirty-cubic-yard bucket is biting into the fossiliferous shales, laying bare a coal seam at the base.

FIGURE 3

The great shovel has swung about and dumped its load of shale. Such heaps of loosened rock are favorite localities for the collector in search of plant fossils.

FIGURE 4

After the coal seam has been uncovered, smaller shovels load the coal directly from the open pit into coal cars.

FIGURE 5

Heavily loaded with its "black gold", a coal train leaves the strip mine, carrying the buried wealth of a distant age to the markets of today.

Photographs by author

more or less isolated and surrounded by protecting deposits of sediment. When split apart, the shaley layers reveal the enclosed fossils in perfect state of preservation.

In northern Illinois, particularly in the area lying in Will and Grundy counties between Morris and Wilmington, a coal vein lies close to the surface, and peculiar conditions of mineralization have preserved the fossils in an unusual state. Here the shale immediately surrounding each individual fossil has become harder than the general mass of sediments, forming rounded or egg-shaped nodules, or concretions. These concretions, sometimes also called iron-stones, contain limited quantities of iron which makes them very hard and gives them a reddish hue after exposure to the weather. Unweathered concretions have the characteristic bluish-grey color of the surrounding shale. When given a violent blow with the hammer, these nodules readily split into two halves revealing the

Photograph by George Langford, Jr.

FIGURE 6
These shale heaps in the Mazon Creek region of Illinois have furnished science with many of the finest plant fossils found anywhere in the world.

Photographs by George Langford

FIGURE 7

Collecting the fossiliferous concretions from the shale heaps in northern Illinois. The fossils within such nodules form natural lines of weakness through the rocks, causing them to split along the plane of the fossils when struck with a hammer.

enclosed fossil which formed a natural line of weakness through the stone. The concretions vary in size from a fraction of an inch to twelve inches or more in length. The average size, however, approximates diameters of five inches in length by two in width. Usually a complete unit of a plant or animal fossil is exposed to view when a concretion is split open.

The usual method of obtaining such fossils is by collecting these concretions from the shale heaps following coal mining operations. The comparatively new method of "strip" coal mining has resulted in the accumulation of vast heaps of shale over wide areas in the mining regions. The mining itself is done by means of enormous electric shovels. First the shale, averaging from twenty to sixty feet in thickness, is scooped away from over the coal seam, exposing it to view in great open pits. The coal is then scooped up by smaller shovels which load it directly into railway cars for transportation. The shale which is shoveled aside in high piles contains the fossiliferous nodules. Having thus been loosened from the general mass of material, the concretions are gathered from the dump heaps by the fossil collector who cracks them open and examines them for fossil contents.

FIGURE 8

An impression of the leaves of *Neuropteris scheuchzeri*. Note the two small orbicular leaflets which characteristically occurred at the base of each larger leaflet in this species. (X ⅔)

ISM 14477

Fossils occurring in shales and concretions such as described above are termed *impressions or incrustrations* (Fig. 8). They are formed in the following way. A branch, leaf, or seed may fall into a nearby stream and drift down toward the sea, lagoon, or lake or a forested area may become flooded. After such a plant part has fallen into the water, the air contained in the plant cells is gradually replaced by water. The fragment thus becomes water-logged and sinks to the bottom.

Sediments carried down by the stream or by a flood of waters may bury the fragment underneath a layer of mud or sand. The entombing material is eventually converted into shale or sandstone, and the enclosed plant becomes a thin coaly layer within.

Sometimes a hollow stem or log becomes filled with mud which assumes the markings on the inside of the woody cylinder. The wood may eventually decay away while the mud hardens into rock, retaining the original pattern. Such fossils are termed *pith-casts* (Fig. 59). The sediments enclosing the outside of a stem or log take on the surface markings of the outer bark, and in the course of time may harden into a *mold.* Meanwhile, the woody material enclosed within the mold may become greatly reduced in volume or may entirely disappear leaving a cavity within the mold which in turn also becomes filled with sediment. This sediment, of course, assumes the surface markings of the mold, and gives rise to a *stem-cast* (Fig. 15). The stem-cast is thus a replica of the original stem, whereas the mold is a reverse of the original stem surface. Both casts and molds, or only one or the other, may remain as a fossil; but in any event, they are all regarded as forms of incrustations.

FIGURE 9

A mummified stem in a lump of coal. University of Chicago Collection. Such forms of preservation within the coal itself are relatively rare among plant fossils. (X ⅓)

Photograph by author

Another type of fossilization is termed *mummification* (Fig. 9). In this case the plant tissues themselves are partially or entirely preserved. This usually occurs when plants are buried in very fine clay which excludes the air and hinders decay. The most common example of mummification is the coal itself, which is a rock composed of plant remains that are more or less impregnated by a fossilized jelly derived from their partial decomposition. As has been previously pointed out, however, the individual plant parts are usually so altered as to make them unrecognizable, although well preserved plant parts may sometimes be seen in the uppermost layers of a coal seam. The best examples of mummification are the insects, flowers, etc. which were imbedded in resin ages ago, the resin since having been converted to amber (Fig. 10).

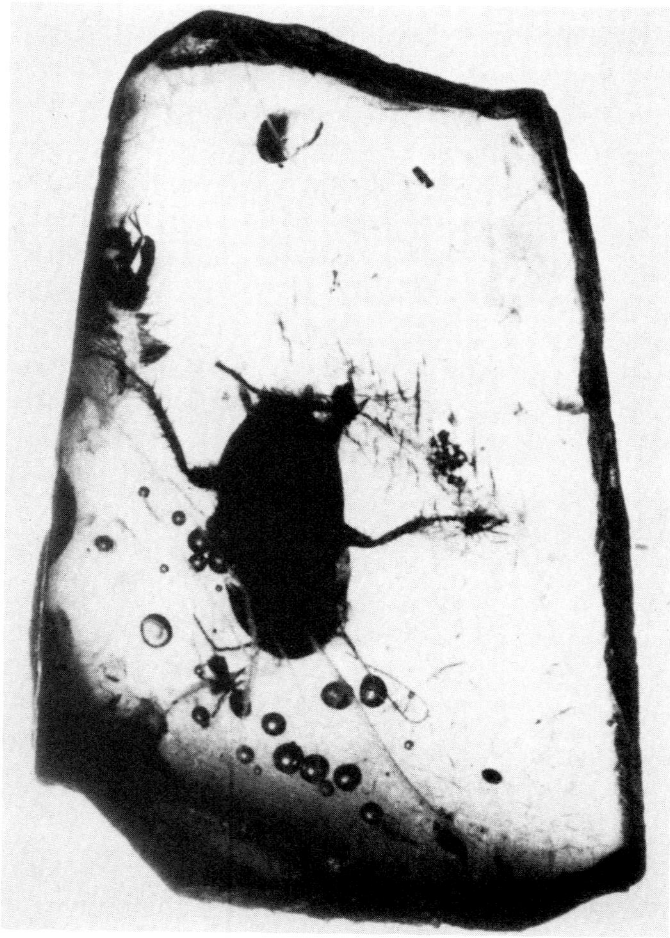

FIGURE 10

The mummy of an insect imbedded in amber. Caught in the sticky resin from a tree millions of years ago, the insect has been hermetically sealed within a permanent transparent case. (X 4)

The term *petrification* is applied when the plant cells have become replaced gradually by minerals deposited from solution along the cell walls, replacing the organic matter bit by bit and sometimes retaining all the details of structure. In this way the organic material of the tissues is replaced by stone. Ordinary petrified wood thus shows the original structure of cells, rays, and growth rings whereas an impression shows only the external form of the bark or internal features of a hollowed stem. It is possible in some instances, of course, for a fossil to be regarded as both a petrification and an impression. Impressions, or incrustations, provide data as to the form of the plant, and petrifications as to its structure.

Fossilization may also occur by the direct preservation of hard parts. Bones, teeth, and shells of animals provide the best known examples, although simple plants such as calcareous algae and siliceous valves of diatoms also may be thus preserved.

All forms of fossilization occur in the plant kingdom, but the impressions are most common, and it is with these that we are here primarily concerned. Their story, written in the rocks, has enabled us to reconstruct with surprising accuracy the types of plants which clothed the lands during the great coal age.

When dealing with living plants we can study all of their parts together, but in working with fossils it is unusual to find complete plants. The stems, roots, leaves, and reproductive organs are generally isolated and separated from each other. Since it is sometimes impossible to piece these parts together and arrive at a reconstruction of the whole, the practice has necessarily arisen of giving distinct generic names to the various parts found. Thus organs with different names may have really belonged to the same plant. Similarly, a genus may also include remains which appear identical but actually belonged to different plants. For example, the genus *Stigmaria* was proposed by the French paleobotanist, Brongniart, early in the nineteenth century for a commonly occurring stem-like impression characterized by numerous rounded scars upon its surface (Fig. 52). He thought that the fossil represented the stems of a distinct type of tree and that the scars marked the attachment places of fallen leaves. It was not until the latter part of the same century that fossil discoveries were made which showed that *Stigmaria* was not a distinct type of tree, but that it was actually the root of various genera of trees. Its surface scars marked the location of former rootlets, but not of leaves as originally thought. Only when *Stigmaria* is actually found associated with the fossil stumps can we be sure as to which type of tree it belonged. It is necessary to retain the use of the generic name, *Stigmaria,* to cover the great numbers of tree roots which are found as detached fragments and cannot be referred to the exact type of tree of which they were once a part. Also, many kinds of seeds occur amid the coal fossils, and they are usually preserved like so many individual detached nuts. Unless actually attached to recognizable portions of stems or leaves, one cannot be sure as to the exact plant upon which such seeds grew. Names given under such conditions are called "artificial form names," as contrasted with true botanical names.

As the study of fossil plants continues, rare specimens are discovered from time to time which settle such pertinent points, and it is then possible to reclassify fossils previously described under form names into more accurate botanical names. Only after the external features, the internal structure, and the reproductive organs are known is it possible to compare a fossil plant with a living type or to place it in a satisfactory phylogenetic classification. The paleobotanist who has discovered and named a new kind of stem or leaf may never know what the entire plant was like, for decades—or even centuries—may elapse before the remainder of the plant becomes fully known. A case in point is that of the common living mulberry which may bear as many as a half dozen varied forms of leaves upon the same twig (Fig. 11).

If such leaves were found separately as fossils, they might well be considered as different species, and only the finding of an entire stem with the several forms attached in place would establish them as actually belonging to the same tree. Having thus established the proof, the scientist is privileged to discard the previously used form names of the fossils and to combine them under a single name. In many respects, the work of the paleobotanist is thus made all the more fascinating, for he never knows when the strike of his pick or the

Photograph by author

FIGURE 11

A twig from the common mulberry illustrates the problems confronting the paleobotanist. Unless found in actual connection fossil leaves of such diverse forms might be easily mistaken for different species. (X ¼)

blow of his hammer may uncover a specimen which will settle a century-old problem and throw new light upon the course which plants have followed in their evolution from simple forms of the past to the complex and magnificent forms of today.

The names of fossil specimens consist of two parts, the generic name and the trivial name. In this respect, they are like our own names, such as John Smith. However, the order is reversed; the generic, or group name coming first, and the trivial, or individual name, second. The two together make up the name of the species. This is followed by the name of the author who first described and named the species, for example, *Lepidodendron lanceolatum* Lesquereux, a common tree of the coal age. When a species has been transferred to a genus other than that to which it was assigned by the original author, the name of the former genus and that of the author are sometimes inclosed in brackets, e. g. *Crossotheca (Staphylopteris) sagittata* (Lesquereux), and sometimes even this may be followed by the name of the person making the change. Because of varying practices in different places and the cumbersomeness of such long names, recent writers have tended to discontinue such designations, using only the name of the original author in all cases. The latter practice is being followed in this handbook.

For the benefit of readers who are not fully acquainted with the systematic botanical classification, a brief outline of the several plant groups is advisable for a better understanding of the fossil plants later to be considered. The plant kingdom is divided into four principal groups, the *Thallophytes, Bryophytes, Pteridophytes,* and *Spermatophytes.*

The *Thallophytes* constitute the simplest of the four divisions. A thallus is a plant showing very little or no differentiation of plant organs. Hence stems, leaves, roots, etc., as such, do not occur among them although some of the more compound forms may have parts analogous to such organs. The sexual organs, when present, usually consist of simple, more or less modified cells, and these may be distributed over the entire plant instead of being restricted locally as in the higher plants. Since the *Thallophytes* are lacking in vascular tissue for conductive purposes, they are ordinarily restricted to wet or moist habitats. The group is divided into three sub-groups constituting the *Algae, Fungi,* and *Bacteria.* All three groups occur as fossils, but are not of sufficient importance to warrant further consideration in dealing with the plant fossils of Illinois. It should be remembered, however, that as agents of decay, bacteria and fungi played an important part in the formation of the gelatinous mummified material which later became coal.

The second great division, the *Bryophytes,* includes two sub-groups, the *Musci* (mosses) and the *Hepaticae* (liverworts). Since these plants have no woody tissue, they also commonly inhabit moist places. The *Bryophytes* are characterized by a distinct alternation of generations in which a generation of sexless spore-producing individuals, called sporophytes, is followed by a generation of egg- and sperm-producing individuals, called gametophytes. The chief distinction between this group and higher plants lies in the relative importance of the two generations. In *Bryophytes,* the so-called "plant" is of the gametophyte generation, while in higher groups it is the sporophyte. *Bryophytes* are quite rare as fossils. Most of the so-called fossil mosses, such as moss-agates, are not plant remains, but are dendritic forms of the mineral manganese which simulates the feathery appearance of mosses.

The third group, the *Pteridophytes,* comprised a large part of the coal age flora, and these are very important as fossils. They are the simplest group possessing vascular tissue whereby water and dissolved chemical matter is transported between the roots and leaves. As in the *Bryophytes,* there is a distinct alternation of generations. *Pteridophytes* reproduce by means of unicellular spores, produced in cases called sporangia. These, in turn, may be borne directly upon the undersides of the leaves or in cones, called strobiles, at the tips of the branches. Some extinct forms bore spore-cases which resembled seeds in general appearance. Modern forms include the club-mosses *(Lycopodium),* horsetails *(Equisetum),* and ferns *(Filicales).* Numerous extinct forms are included in the group, and these will be considered fully in later sections of the handbook.

The *Spermatophytes* include all of the higher plants bearing true seeds. The group is divided into two large sub-divisions, the *Gymnosperms* and the *Angiosperms.* The former are characterized by their habit of bearing the seeds in a naked condition, whereas the *Angiosperms* produce their seeds within a closed structure called the ovary. The latter include most of the important plant forms of the present day, such as the grasses and grains, the broad-leaved fruit trees, and the food-bearing vegetables and herbs. These did not evolve until millions of years after the close of the coal age, and since rocks bearing their fossils do not occur in Illinois, they will not be considered here. The *Gymnosperms,* on the other hand, comprising many living and extinct forms, including the cycads, the maidenhair or Ginkgo tree, and the evergreen conifers, were important members of the coal age flora, and will be considered in detail. Much of the fern-life foliage previously linked with the ferns has been found to belong to members of this group.

Exceptions and confounding perplexities are frequently encountered in nature. Orders, families, genera, and species are merely collections of plants that have greater resemblances toward each other than they have differences. Individual specimens, both fossil and living, may be found that only partially conform to the characteristics of a given group and tend to grade into other groups. Allowances must be made for such exceptions, and care must be taken when identifying specimens to avoid confusion caused by the deceptive resemblance of some character due to adverse conditions of preservation.

For the reader's convenience, all of the form genera included in this volume are shown in the following classification table. Because fern-like types cannot be differentiated solely upon the basis of their form, the ferns and seed-ferns are classified as a unit under the *Pteridophytes.*

Group—Pteridophyta

Order—Lycopodiales

Family—Lepidodendreae

Genera: Asolanus, Aspidiaria, Aspidiopsis, Bergeria, Knorria, Lepidodendron, Lepidophloios, Lepidophyllum, Lepidostrobus, Lycopodites, Omphalophloios, Sigillaria, Sigillariophyllum, Sigillariostrobus, Stigmaria, Syringodendron, Ulodendron

Order—Equisetales

Family—Calamarieae

Genera: Annularia, Asterophyllites, Calamites, Calamophloios, Calamophyllites, Calamostachys, Dictyocalamites, Macrostachya, Paleostachya, Pinnularia, Volkmania

Order—Sphenophyllales

Family—Sphenophylleae

Genera: Sphenophyllum, Sphenophyllostachys

Orders—Filicales and Cycadofilicales

Family—Sphenopterideae

Genera: Aphlebia, Corynepteris, Diplothmema, Eremopteris, Renaultia, Sphenopteris

Family—Pecopterideae

Genera: Acitheca, Asterotheca, Crossotheca, Dactylotheca, Dicksonites, Pecopteris, Ptychocarpus, Senftenbergia

Family—Mariopterideae
Genus: Mariopteris

Family—Alethopterideae
Genera: Alethopteris
Desmopteris

Family—Neuropterideae
Genera: Cyclopteris, Linopteris, Mixoneura, Neuropteris

Family—Odontopterideae
Genus: Odontopteris

Incerta sedis
Genera: Caulopteris
Megaphyton
Pinnularia
Ptychopteris
Spiropteris

Group—Spermatophyta
Sub-group—Gymnospermae
Order—Cordaitales
Family—Cordaiteae
Genera: Artisia, Cordaianthus, Cordaicarpus, Cordaicladus, Cordaites, Dorycordaites, Eucordaites, Poacordaites

Gymnospermous Seeds
Genera: Carpolithus, Codonospermum, Codonotheca, Cordaicarpus, Holcospermum, Lagenospermum, Pachytesta, Rhabdocarpus, Samaropsis, Trigonocarpus

CHAPTER II

Lycopodiales

THE *Lycopodiales* is an order of plants belonging to the *Pteridophyte* group. The order during the coal age included a much larger, more numerous, and more highly differentiated variety of forms than exist today. The majority of the living forms are found growing as small terrestrial plants or as epiphytes in moist tropical forests. A few species are found in northern regions. The two most commonly known genera are *Selaginella,* a club-moss, and *Lycopodium,* the diminutive ground-pine used in the making of Christmas wreaths.

During the coal age the *Lycopodiales* were mostly arborescent forms equal to, or exceeding in size, the average forest trees of today. They were extremely abundant, had a world-wide distribution, and constituted an important element in the formation of coal. Some coals, such as cannel coals, seem to be made up almost entirely of the spores of these trees.

Lepidodendron Sternberg (Fig. 12) is a fossil genus which includes trees similar in form and habit to the existing *Lycopodium* except for size. The living form is an insignificant dwarf in comparison to its gigantic ancestors. The trunks of these fossil trees tapered gradually, reaching heights of a hundred feet or more, and diameters up to two feet. The trees grew tall and straight for a considerable distance above the ground, whereafter the trunks branched dichotomously, that is, by twos—each fork dividing again and again to the ultimate branch tips. The forking of the branches, as in modern trees, often developed unequally, however. The bark of the trunk and branches of *Lepidodendron* is characterized by a diamond-shaped pattern in which each unit represents the former point of attachment of a fallen leaf. The name, which means "scale-like tree," refers to this pattern. Pecu-

FIGURE 12

Restoration of a *Lepidodendron* tree. The diamond-shaped patterns of the bark represent the attachment places of former leaves. These scars continued to grow in size even though the leaves had fallen.

Courtesy Chicago Natural History Museum

liarly enough, these leaf cushions continued to grow even though the leaves themselves had long since fallen off. The younger branches bore spirals of linear or needle-like leaves which persisted for a time and then were shed. The leaves seem to have been clustered about the cones, or spore-bearing strobiles, which were produced on the tips of the young branches (Fig. 13). The bark on the mature trunk and stems of *Lepidodendron* thus consisted of diamond-shaped leaf cushions placed adjacent to each other, each of which had near its center a rhomboidal leaf scar representing the actual attachment point of the former leaf. On well preserved specimens, three dot-like marks may be seen near the middle

FIGURE 13

The cones of the *Lepidodendron* trees occurred at the tips of the branches, and were surrounded by a cluster of grass-like leaves which persisted for a short distance down the stem. (X 1/6)

Photograph by the author from a restoration at the Chicago Natural History Museum.

of the leaf scar (Fig. 14). The central dot represents the point at which the vein, or vascular bundle, passed from the stem into the leaf. The two lateral dots are believed to represent a strand of aerating or secretory tissue, called parichnos

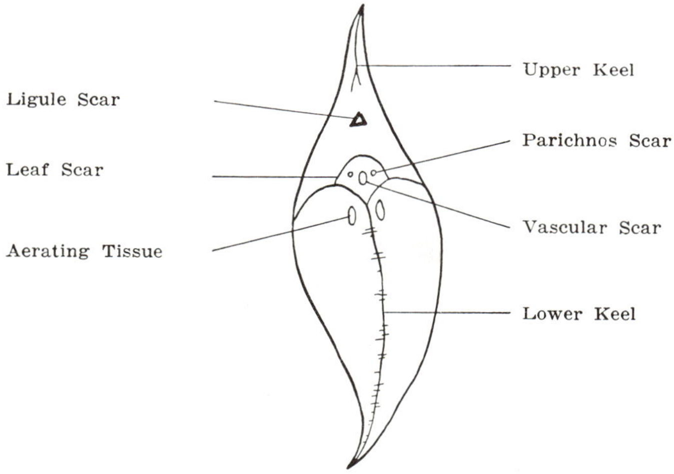

FIGURE 14

Diagram of a single leaf cushion of *Lepidodendron aculeatum.* (Slightly enlarged)

tissue. During life, this tissue had an outlet to the air through two other small dot-like openings which may be seen just beneath the leaf scar. Extending above and below the leaf scar was usually a central ridge called a keel. In certain species it bears characteristic transverse wrinkles or markings. In some species a small triangular mark may be seen slightly above the center of the leaf scar. It is called the ligule

Photograph by author

FIGURE 15
Lepidodendron aculeatum Sternberg. (X 1) From a coal mine in southern Illinois, exact locality unknown.

scar and represents the pit in which the ligule, a small outgrowth from the base of the leaf, was seated. As in modern plants, its function is believed to have been for the purpose of preventing the water which drains down the leaf from collecting in the leaf axis.

Lepidodendron aculeatum Sternberg (Fig. 15) is a common Illinois species. The leaf cushions are at least three times as long as their greatest width and curve in opposite directions at each end. The leaf scar occurs about two-thirds of the way up from the bottom of the cushion and is at least as high as it is wide. Two lines descend from the angles of the leaf scar and join the margins of the cushion near its midpoint. The ligule scar is visible on well-preserved specimens just above the leaf scar. The upper keel begins slightly above

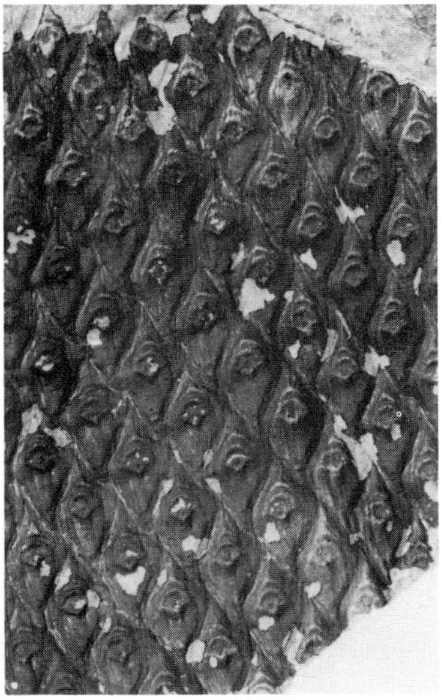

FIGURE 16
Lepidodendron obovatum Sternberg. (X 3/5) Collected in Illinois, exact locality unknown.
ISM 416532

the ligule scar and continues to the apex, thereby extending for only about half the distance of the upper portion of the cushion. The lower keel extends throughout the entire distance of the lower part of the cushion, and it is usually marked by numerous transverse wrinkles.

Lepidodendron obovatum Sternberg (Fig. 16) is quite similar to *L. aculeatum,* but in this species both the leaf cushions and the leaf scars are wider in proportion to their height, the length being not more than twice their width.

FIGURE 17 (Right)

Lepidodendron lanceolatum Lesquereux. (X 1)

•

FIGURE 18 (Below)

Lepidodendron loricatum Arber (X ¾)

ISM 14382

Courtesy Illinois State Geological Survey

Also, both ends are obtusely terminated. The upper keel is less distinct than in *L. aculeatum,* and the transverse wrinkles on the lower keel are fewer and fainter, or sometimes absent.

Lepidodendron loricatum Arber (Fig. 18) has been united by some paleobotanists with *L. obovatum.* There are, however, several features which tend to distinguish it as a separate species. Its leaf scar is characteristically more rhomboidal in shape than is that of *L. obovatum* and is situated higher up on the cushion. In addition, the lines descending from the angles of the leaf scar join the margin of the cushion at a higher point and at a lesser angle than do those of *L. obovatum.*

Lepidodendron lanceolatum Lesquereux (Fig. 17) has leaf cushions that are long and narrow with somewhat the same proportions as *L. aculeatum* except that they are

FIGURE 19

Lepidodendron wortheni Lesquereux. (X 5/3) Collected near Murphysboro, Illinois. ISM 1730

FIGURE 20 (Right)

Lepidodendron veltheimi Sternberg. (X 1)

ISM 14338

FIGURE 21 (Below)

Lepidodendron rimosum Sternberg. Has wrinkled bands of cortex between the leaf cushions (X 1)

ISM 14770

Photograph by George Langford

straighter vertically. The leaf scar, however, is narrower and is characteristically notched on the upper side. The ligule scar is usually quite distinct. The upper keel is frequently indistinct, but the lower one is quite clear. It, however, bears no transverse markings.

Lepidodendron wortheni Lesquereux (Fig. 19) has leaf cushions which are longitudinally unsymmetrical, and with considerable variation between specimens. An elliptical leaf scar occupies the entire width of the cushion, and the margins are not always clearly defined. There are no keels, but numerous transverse wrinkles occur in the cushion both above and below the leaf scar.

Lepidodendron veltheimi Sternberg (Fig. 20) is easily recognized by the fact that the leaf cushions, which are about three times as high as wide, are not set directly against each other but are separated by wide bands of smooth cortex. The cushion has a lower keel which is crossed by faint transverse wrinkles.

Lepidodendron rimosum Sternberg (Fig. 21) is quite similar to *L. veltheimi* except that the separating bands of cortex between the cushions contain numerous oblique wrinkles or ridges.

Lepidodendron rigens Lesquereux (Fig. 22) includes top branches with long linear leaves. The leaf cushions on the young branches do not yet have their final form, and since the leaves themselves are not diagnostic, the branches cannot be assigned to definite mature species. The leaf cushions are rhomboidal, the leaf scars are of the same form, but are shorter, covering the upper half of the cushions.

Omphalophloios White is a genus closely allied to *Lepidodendron.* The leaf cushion has the characteristic diamond-shaped aspect of *Lepidodendron,* but the leaf scar has a subtriangular or circular shape as contrasted to the distinctly rhomboidal scar of the other genus.

Omphalophloios cyclostigma Lesquereux (Fig. 23) has leaf cushions destitute of keels or ornamentation. The

Courtesy Illinois State Geological Survey

FIGURE 22

Lepidodendron rigens Lesquereux.
(X 1)

boundary of the rounded or subtriangular leaf scar is frequently a sharp, clear furrow.

Asolanus Wood is a distinct genus also closely related to *Lepidodendron*. As yet, only one species has been recognized,

FIGURE 23 ISM 14497

Omphalophloios cyclostigma Lesquereux.
(X 4/5)

Asolanus camptotaenia Wood (Fig. 24). It is characterized by distant spirally arranged leaf scars between which the bark is marked with oblique lines which radiate outwardly in a general way from the scars. One or two tiny dot-like marks may be seen in the center of the scars.

Lycopodites Goldenberg was a small herbaceous lycopod which most nearly resembled the modern ground-pines of the present day. The stems bore tiny needle-shaped leaves which may very easily be mistaken for the leafy twigs of some *Lepidodendrons*.

FIGURE 24

Asolanus camptotaenia Wood.
Right: (X 1) ISM 14284
Left: (X ½)

← Photograph by George Langford

Lycopodites meeki Lesquereux (Fig. 25) may be recognized by its slender stems and branches which fork dichotomously. The very small leaves are closely overlapping upon each other and contain no veins. It bears a close resemblance to its modern descendant *Lycopodium*.

Lepidophloios Sternberg is a comparatively rare genus, of which numerous specimens have been found in Illinois. Like *Lepidodendron*, it consists of tree forms with forking branches bearing rhomboidal leaf cushions arranged in a spiral manner. In this genus, however, the cushions are always broader than high, whereas the reverse condition is true in *Lepidodendron*. Also, in *Lepidophloios* the leaf scar normally occurs at, or very near, the top of the cushion while in the other genus the scar is slightly above the center of the

FIGURE 25
Lycopodites meeki Lesquereux. (X 1)
ISM 14418

cushion. Sometimes, in the older stems, the leaf scar of *Lepidophloios* may be at the very bottom of the leaf cushion, and the cushions themselves may overlap one another. These phenomena are probably the result of varying growth conditions, or the collapse of underlying tissues.

Lepidophloios laricinus Sternberg (Fig. 26) is the most common Illinois species. The leaf cushion is two to three times as wide as high and is acutely pointed at the sides. The leaf scar is proportionately of the same shape as the cushion. It may be located either at the extreme top or bottom of the cushion, and the three dot-like vascular marks

FIGURE 26

Lepidophloios laricinus Sternberg. (X 1)
From a coal mine at Danville, Illinois.

Photograph by author

Photograph by George Langford

FIGURE 27

Lepidophloios protuberans Lesquereux. (X 1½)
ISM 14771

may be seen near the center. Slight indications of a keel may be present in some specimens.

Lepidophloios protuberans Lesquereux (Fig. 27) is characterized by leaf cushions that are somewhat distant, being separated from each other by undulating continuous wrinkles. The cushions are somewhat ovate in form, are enlarged below the middle, and then are rather abruptly narrowed to a truncated basal point. The leaf scars, located at

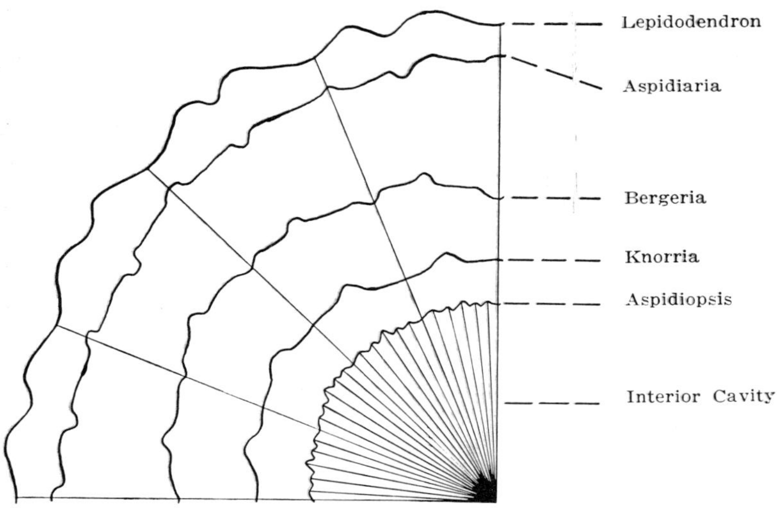

FIGURE 28

Diagram of cross-section of a *Lepidodendron* trunk showing relationship of the various bark layers and central area. (After Susta.)

the top of the cushions, are rhomboidal with the lower angle distinctly rounded and the lateral angles acute.

Before their true nature was understood, imperfectly preserved stems and trunks of *Lepidodendron* and related genera were thought to be different types of trees, and their impressions were classified accordingly. *Aspidiaria* Presl, *Bergeria* Sternberg, *Knorria* Sternberg, and *Aspidiopsis* Potonié are actually different layers of the same tree (Fig. 28). *Lepidodendron* is, of course, the epidermis or outer bark of the tree. *Aspidiaria* (Fig. 29) is the outer cortex, or the first layer beneath the surface bark. *Bergeria* (Fig. 30) represents the middle cortex, and *Knorria* (Fig. 31) is the inner cortex of the stem. *Aspidiopsis* is an impression of the interior cavity of the trunk. All transitions may be found between these layers due to the state of decomposition which

FIGURE 29

Aspidiaria, or sub-surface bark of a *Lepidodendron* trunk. (X ¾)
ISM 14286

had been reached before fossilization took place. Such impressions may vary greatly in appearance from specimen to specimen. Some attempts have been made to classify them into species, but it is of little use to do so since they have no botanical or geological value.

FIGURE 30

Bergeria or middle bark layer of a *Lepidodendron* trunk. (X 1/5) ISM 14751

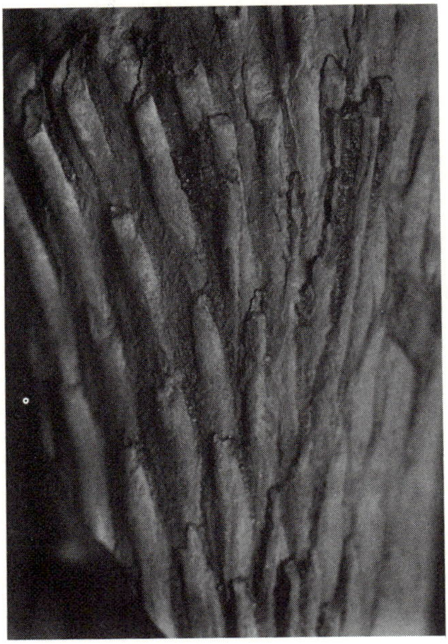

Photograph by author

FIGURE 31

Knorria, or inner bark of a *Lepidodendron* trunk. (X 1) Collected near Danville, Illinois.

Ulodendron Lindley & Hutton (Fig. 32) is a genus characterized by the presence of large umbilical depressions varying from an inch to several inches across and surrounded, in different specimens, by various types of leaf cushions. They represent knots left by fallen branches and should be classified according to the tree upon which they occurred. The generic name is still useful, however, for application to specimens on which the leaf cushions are indistinct and cannot properly be assigned to the parent tree.

FIGURE 32

Ulodendron, or branch scar on the trunk of a *Lepidophloios* tree. (X 5/8)

ISM 14559

Sigillaria Brongniart (Fig. 33) was a tree of wide distribution during the coal period, but was less common numerically than *Lepidodendron*. In contrast to the latter it seldom, if ever, bore branches. In this respect it more nearly resembled the modern palm tree in general aspect. It attained heights in excess of a hundred feet, with diameters up to six feet. The straight, gently tapering trunk was surmounted by a cluster of long grass-like leaves which are

Courtesy Chicago Natural History Museum

FIGURE 33

Restoration of a *Sigillaria* tree. The tall unbranched trunks were usually surmounted by a crown of very long, narrow leaves.

usually indistinguishable from those of *Lepidodendron* or *Lepidophloios*. After the leaves had dropped from the trees, a series of leaf scars, usually hexagonal in general outline, became visible on the bark. In most cases there were no true leaf cushions. The scars occur directly above one another in vertical rows, and each row is separated by a straight or undulating furrow extending the full length of the trunk. In some forms the scars are situated very close together and care must be taken not to confuse such species with *Lepidophloios;* other species have the scars widely separated vertically and

Photograph by author

FIGURE 34

Sigillaria mammillaris Brongniart. (X 1) From a coal mine at Christopher, Illinois.

horizontally, and the bark between may be smooth or bear characteristic wrinkles or markings. The reproductive spores were borne in cones supported on small stems which emerged directly from the tree trunk.

Sigillaria mammillaris Brongniart (Fig. 34) is characterized by prominent closely placed leaf scars which occupy the whole breadth of the rib between the furrows. The middle portion of the scar may be slightly wider than the upper and lower portions thus giving the furrow between them a slightly undulating appearance. The lines descending from the upper lateral angles of the scar merge into the furrow at the widest point. Transverse wrinkles sometimes extend across the narrower portion of the rib between each scar.

FIGURE 35

Diagram of *Sigillaria elegans* Sternberg, showing the tile-like hexagonal leaf scar characteristic of this species. (X 1)

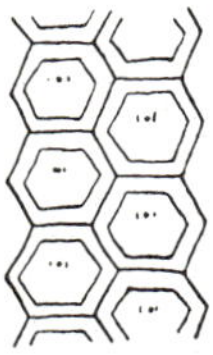

Sigillaria elegans Sternberg (Fig. 35) has its scars placed directly adjacent to each other, and the vertical furrow between them is an angular zig-zag line. Each scar is a nearly equal-sided hexagon so that the upper and lower terminus of each is a straight horizontal line. The bark pattern as a whole thus resembles the design of tile in a bathroom floor.

FIGURE 36

Sigillaria tessellata Brongniart. (X 1) Collected at Mariemont, France. (From Renier)

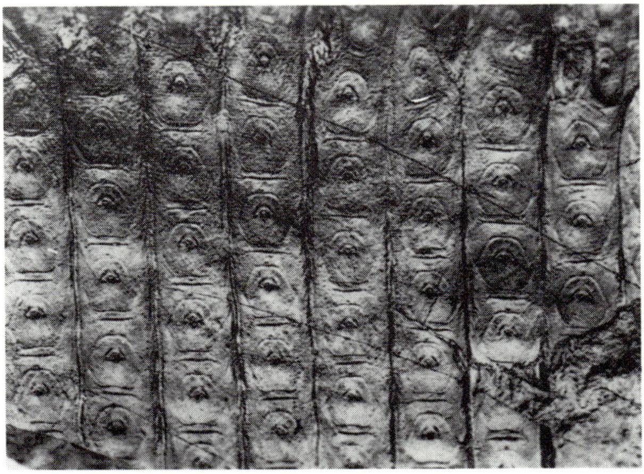

The three dot-like marks, representing the former attachment of the vascular strands, are usually distinctly visible slightly above the center of the scar.

Sigillaria tessellata Brongniart (Fig. 36) differs from *S. mammillaris* by its straight unornamented ribs separated by straight vertical furrows which may be emphasized by numerous short oblique wrinkles. The leaf scars occupy only about two-thirds of the width of the ribs, and that portion of the rib between the scars and furrows is quite smooth. A prominent wrinkle often exists immediately beneath the leaf scar, but it does not extend across the entire distance between the furrows. The lower width of the leaf scar is slightly greater than the upper width.

FIGURE 37

Sigillaria rugosa Brongniart. (X 1) Collected at Courcelles-Nord, France. (From Renier)

Photograph by author

FIGURE 38

Sigillaria scutellata Brongniart. (X 1) Collected in Illinois; exact locality unknown.

Sigillaria rugosa Brongniart (Fig. 37) is characterized by a great vertical interval between scars. The distance by which a scar is separated from the next one above or below it amounts to approximately four or five times the height of the individual scars. The lines descending from the lateral angles of the leaf scar do not diverge into the furrows, but extend downward below the scar and gradually fade out. The area enclosed between these descending lines is quite rough, or rugose, and continues in lesser degree toward the next scar below. Immediately above the scar, however, the rugosity is again more pronounced. It is this feature which has given the name to the species.

FIGURE 39

Sigillaria laevigata Brongniart. (X ¾) Collected in Johnson Co., Illinois.

ISM 416531

Sigillaria scutellata Brongniart (Fig. 38) differs from *S. rugosa* in the closer spacing of its scars and the absence of rugosity. The vertical distance between the scars is only about twice their individual height. Also, the descending lines from the leaf scars emerge at a wider angle and run into the furrows.

Sigillaria laevigata Brongniart (Fig. 39) has wide ribs separated by prominent furrows. The leaf scars are hexagonal in form with the top border straight or slightly notched. Frequently the three lower sides are indistinct, thus giving the appearance of a continuous arc or oval. The proportionate sizes of the ribs and leaf scars is a marked feature of this species. The large ribs measure one and a half to two inches

Photograph by F. C. MacKnight

FIGURE 40
Sigillaria orbicularis Brongniart. Note that the surface bark has been removed at the left, exposing the under layer, usually called *Syringodendron*. Compare with Fig. 42 From Danville, Illinois. (X 1)

in width, while the leaf scars do not exceed three-quarters of an inch in width. The vertical distance between the scars is about equal to the height of the individual scars. The entire rib may be covered with fine vertical striations.

Sigillaria orbicularis Brongniart (Fig. 40), as its name indicates, is characterized by scars which are very nearly round. These are separated vertically from each other by an interval which is approximately one and one-half times the height of the individual scars.

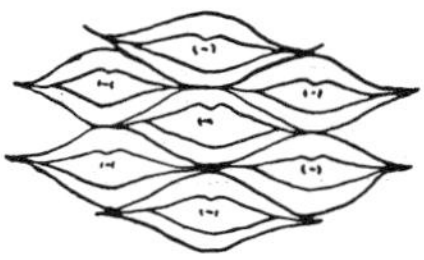

FIGURE 41

Diagram of *Sigillaria brardi* Brongniart, showing characteristics of the leaf cushions and scars. (X 1)

Sigillaria brardi Brongniart (Fig. 41) is somewhat similar in general aspect to *Lepidophloios*. Both cushions and leaf scars are present, the scars occupying the central part of the

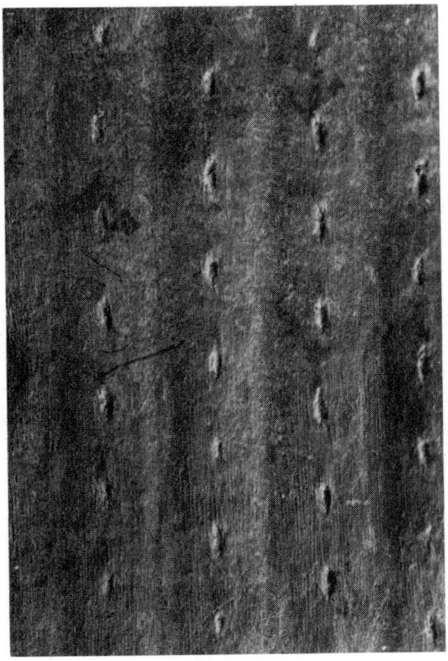

Photograph by author

FIGURE 42

Syringodendron or sub-surface bark of a *Sigillaria* tree. (X 1)

cushion. The upper and lower borders of both are rounded, while the lateral borders are elongated into very acute angles.

A horizontal line extending from each side of the leaf scar unites it with the cushion. In this respect the species differs from *Lepidophloios,* wherein the scar is located either at the extreme top or bottom of the cushion. The cushions are situated directly adjacent to each other so that the sharply

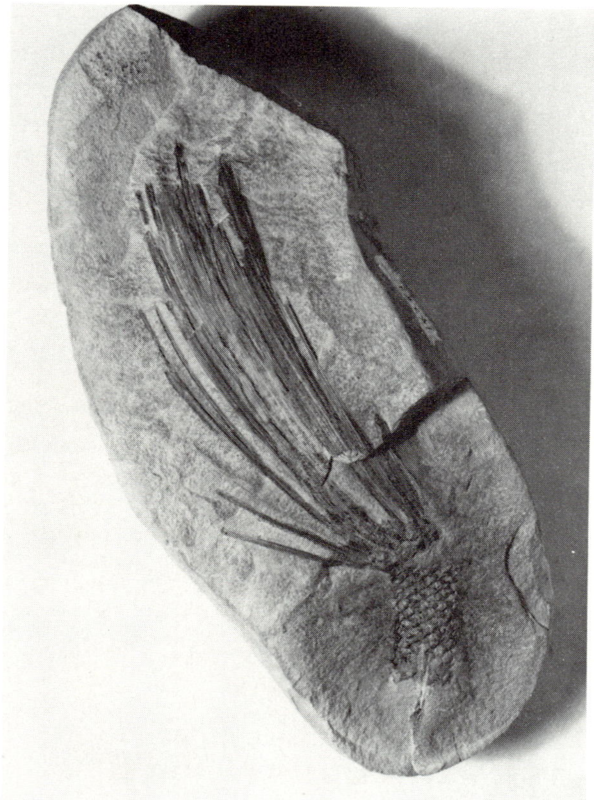

FIGURE 43

Lepidophyllum, or leaves of the *Lepidodendron* tree, in this case of *L. obovatum* Sternberg. (X ½)

ISM 14383

zig-zag furrow between them is not so prominent as in other species.

As in the case of *Lepidodendron,* sub-surface layers of *Sigillaria* stems and trunks are also found as fossils. These are usually referred to the "genus" *Syringodendron* Sternberg (Fig. 42). Since they merely represent decorticated portions of well known trees, they have little practical value.

Lepidophyllum Brongniart and *Sigillariophyllum* Grand 'Eury include leaves, both sterile and fertile, which grew on the lycopod trees. When leaves are found attached to definitely recognizable stems, they can, of course, be assigned to their proper genus (Fig. 43). However, since the leaves of many lycopod trees were very similar, detached specimens

FIGURE 44

Lepidophyllum longifolium Brongniart. (X ½)

ISM 14398

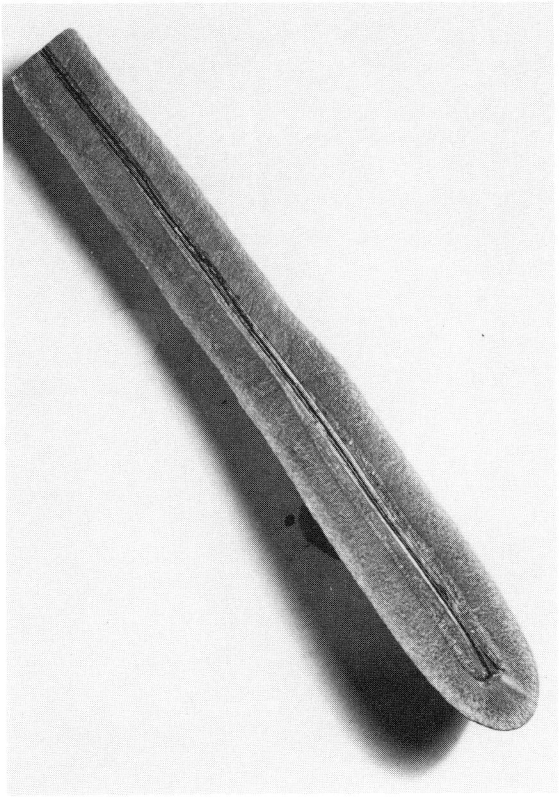

cannot be assigned with certainty to definitely known parent plants. Because the *Lepidodendron* trees were more numerous than the other forms, all leaves which cannot be classified otherwise are referred to the "form genus," *Lepidophyllum*. Some forms were short and more or less lanceolate or arrow-shaped. Others were long and grass-like attaining lengths of two feet or more. All types possessed a single vein extending down the center for their entire length. The so-called fertile leaves were usually short, triangular or spear-shaped forms which grew on the cones and served as sort of scale-like protections to the spore cases. The longer grass-like leaves represented the leafy foliage on the stems of the trees.

FIGURE 45

Lepidophyllum majus Brongniart. (X 1)

ISM 14772

Photograph by George Langford

Lepidophyllum longifolium Brongniart (Fig. 44) includes the long grass-like blades which constituted the foliage of the lycopod trees. They average about one-quarter of an inch in width, and are usually from one to two feet in length. When found associated with stems bearing identifiable leaf cushions, they are classified according to the tree species upon which they grew.

The shorter bract-like leaves, borne on the fertile cones, are classified according to their relative shapes and sizes. *Lepidophyllum majus* Brongniart (Fig. 45) is a large, oblong, acutely pointed, lanceolate leaf. Average specimens are seven to nine centimeters long, and thirteen to sixteen millimeters wide.

FIGURE 46

Lepidophyllum lanceolatum Brongniart. (X 4/5)
ISM 14406

Lepidophyllum lanceolatum Brongniart (Fig. 46) is about half the size of *L. majus*. The leaves are usually about three and a half centimeters long, and seven to eight millimeters wide.

Lepidophyllum intermedium Lindley & Hutton (Fig. 47), as its name indicates, is intermediate in position between *L. majus* and *L. lanceolatum*. The leaves average three to five centimeters in length, by six to nine millimeters in width.

Lepidophyllum lancifolium Brongniart (Fig. 48) is characterized by a slightly different form. The leaves are shorter in relation to their width than in the foregoing species. Their average length is about two and a half centi-

FIGURE 47
Lepidophyllum intermedium Lindley and Hutton. (X ¾)
ISM 14405

meters, with a width of five millimeters. The basal corners are drawn out into sharply pointed flares.

Lepidophyllum ovatifolium Lesquereux (Fig. 49) has a more or less ovate form, as its name indicates. In contrast to

FIGURE 48

Lepidophyllum lancifolium Brongniart. (X 1)
ISM 14413

FIGURE 49

A cluster of *Lepidophyllum ovatifolium* Lesquereux attached to a fragment of *Lepidodendron* cone. (X ¾)
ISM 14411

other forms, the tip is bluntly pointed. The leaves are about one and a half centimeters long, and are widest near the base where they average seven to eight millimeters in width.

Lepidophyllum triangulare Zeiller (Fig. 50) is practically triangular in form. The leaves are twelve to fifteen millimeters in length, and are widest at the base where they measure eignt to twelve millimeters in width.

The lycopod trees reproduced by means of spores borne in cones, or strobiles. These are usually referred to the

FIGURE 50
Lepidophyllum triangulare Zeiller. (X 1)
ISM 14412

"genus" *Lepidostrobus* Brongniart (Fig. 51), although the name *Sigillariostrobus* Schimper is also used for the inclusion of specimens which have unmistakable affinities with the *Sigillarias*. The cones were cylindrical or ovoid, and averaged three to six inches in length by approximately three-quarters of an inch in diameter. Unusually large specimens were a foot or more in length by two inches in diameter. They consisted of a central stalk crowded with spirally arranged bracts, or leaves, the upper surface of each bearing a single sporangium, or spore case. Since secondary wood is absent from the stalks, it is believed that the cones were shed when ripe and that fertilization occurred on the damp swampy ground.

The roots of the lycopod trees are referred to the "genus" *Stigmaria* Brongniart. As has been previously pointed out, these impressions were thought for a long time to be the stems of a distinct tree. Later, stumps of lycopod trees standing

FIGURE 51

A cross-section through *Lepidostrobus,* with cone surrounded by attached leaves. (X 1) Collected by the Illinois State Geological Survey.

ISM 1622

in situ where they once grew were uncovered which showed the stigmaroid roots extending downward into the former soil. The root system of the trees consisted of four main branches, each of which dichotomized several times to form a ramified network in the soil or humus formed by the decay of previous generations of swampland plants. Numerous lateral rootlets, up to two feet or more in length, occurred at closely spaced intervals along the main roots.

Nearly all of the roots are referred to *Stigmaria ficoides* Sternberg (Fig. 52). Irregular spirals of circular scars appear on the surfaces, marking the attachment points of the former rootlets. A central pit in each scar indicates the position of the vascular bundle. Between these rootlet scars pass longitudinal undulating lines or wrinkles.

FIGURE 52

Stigmaria ficoides Sternberg, or root of a lycopod tree. The circular scars mark the attachment places of former lateral rootlets. Collected in southern Illinois; exact locality unknown. (X ½)

ISM 1631

Several other species of *Stigmaria* have been named by early paleobotanists, based largely on the sizes of the rootlet scars and the character of the wrinkles. These species are of doubtful value since they cannot be linked with their parent trees, and at best may represent nothing more than individual variations in size or different modes of preservation.

CHAPTER III

CORDAITALES

IN ADDITION to the various forms of lycopod trees, the coal forests of Illinois contained gymnospermous trees belonging to the higher order *Cordaitales*. These trees were less common than the lycopods, but nevertheless were

FIGURE 53

Restoration of a *Cordaites* tree, showing its peculiar strap-shaped leaves.

Courtesy Chicago Natural History Museum

very widely distributed throughout the world. In some foreign regions, however, they seemed to constitute locally the dominant tree growth. The trees attained, or even exceeded, heights of a hundred feet, but did not exceed diameters of two feet. The branches bore large strap-like leaves which were sessile, or stemless, and were spirally arranged

FIGURE 54

Portion of leaf of *Cordaites borassifolius* Sternberg. (X 2/5)
ISM 14330

around the stems. The leaves are classified under the general term *Cordaites* Unger, and the trunks—which are very rare—under the genus *Cordaicladus* Grand 'Eury. It is believed that these trees grew in the highlands away from the coal swamps proper, and that the leaves may have drifted down the streams to become buried and preserved while the tree trunks themselves died and decayed on the uplands.

The leaves of *Cordaites* have been placed into three sub-genera by Grand 'Eury, based on their shapes and sizes. *Eucordaites* includes very large strap-like leaves up to two or three feet in length and several inches wide. The tips of the leaves in this group are always rounded. *Dorycordaites* includes leaves nearly as large, but with pointed tips. *Poacordaites* has relatively small narrow grass-like leaves with obtusely rounded tips.

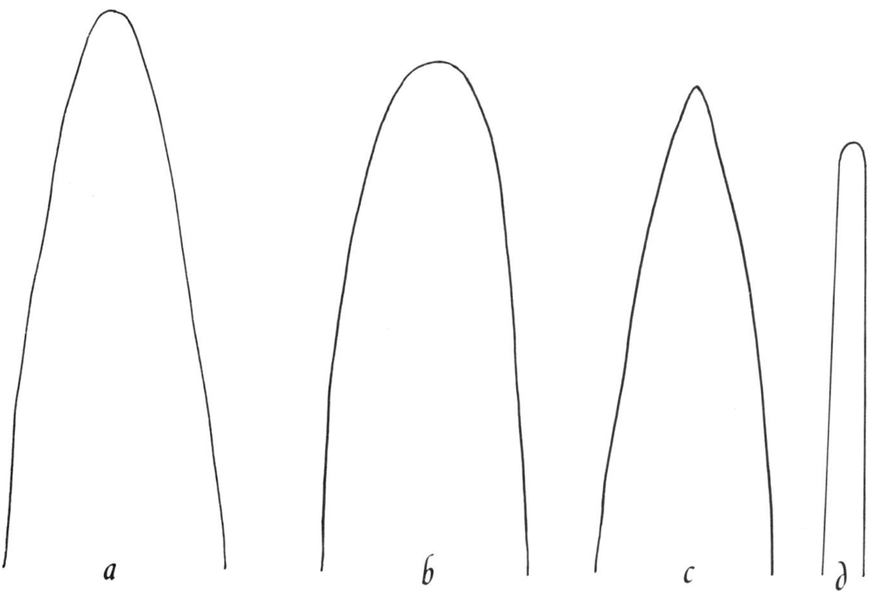

FIGURE 55

Diagram of leaf tips of *Cordaites*. (X 1)

a. *Cordaites borassifolius* Sternberg;
b. *Cordaites principalis* Germar;
c. *Dorycordaites palmaeformis* Goeppert;
d. *Poacordaites microstachys* Goldenberg;

In all species of *Cordaites* there are numerous veins which traverse the leaves longitudinally and rarely fork. Some species are differentiated on characteristics of this venation, but the method is not reliable in all cases because it is frequently impossible to distinguish between true veins and longitudinal strands of tissue. Sometimes too, poor preservation may have obliterated the finer venation.

Cordaites borassifolius Sternberg (Figs. 54 & 55, a) includes leaves of the *Eucordaites* type, being long and strap-like with an obtuse apex. They are usually about two feet long and are widest in the middle where they measure from an inch and a quarter to four inches across. At the base, measurements vary from three-quarters of an inch to two inches in width. The veins, when they can be recognized, are alternately thick and thin.

Photograph by author

FIGURE 56

Artisia, or pith cast of a *Cordaites* tree. (X 1) Collected in Illinois; exact locality unknown.

Cordaites principalis Germar (Fig. 55, b) has leaves which may be a few inches shorter than *C. borassifolius,* and somewhat narrower. The apex is more rounded and obtuse, however. One to five very fine lines occur between each pair of heavier veins. When only small fragments of the leaves are found, it is frequently impossible to tell which of the two species is represented.

Dorycordaites palmaeformis Goeppert (Fig. 55, c) includes linear-lanceolate leaves with pointed tips. They may reach sizes approximating those of *C. borassifolius*. The veins are very fine and are all of equal strength.

Poacordaites microstachys Goldenberg (Fig. 55, d) has grass-like leaves about a foot long by three-eighths of an inch wide, or less. The tips are obtuse, and the veins may all be

FIGURE 57

Cordaianthus, or male catkin of the *Cordaites* tree. (X 1)

ISM 14773

Photograph by George Langford

of equal strength, or with one or two alternating thinner lines between the heavier ones.

As is the case with other trees, pith casts are sometimes found of *Cordaites,* and these are referred to the "genus" *Artisia* Sternberg (Fig. 56). The young stems had a solid pith which did not seem to keep pace with the elongation of the stem. The faster growing stem caused the pith to rupture at intervals, leaving transverse diaphragms at these points, and it is these which may be seen in the fossil impressions.

The fructifications of the *Cordaites* trees consisted of catkins borne on the branches. The male, or staminate, catkins are classified under the name *Cordaianthus* Grand 'Eury (Fig. 57). The female, or ovulate, catkins produced rounded seeds classified as *Cordaicarpus* Geinitz (Figs. 161, d & 164). The seeds were attached in the axils of the bracts which were spirally arranged around the axis of the catkins. Narrow wings encircled the seeds, which may have been an aid in wind dispersal.

CHAPTER IV

EQUISETALES

THE *Equisetales* was one of the most widely distributed orders of plants during the Pennsylvanian coal-forming period. Although many of them attained the proportions of trees, actually they were gigantic rushes. They lined the streams and water courses of their time just as the bamboos and bullrushes do today. These latter, however, are not members of the same group. The more common Carboniferous forms attained heights of fifty feet, and diameters of two feet or more. From thick underground stems, or rhizomes, arose trunks which in turn gave rise to whorl upon whorl of successively smaller branches. These trunks were jointed in appearance, some types bearing branches at every joint, or node, and others only at distant intervals. The leaves were always borne in rings, or whorls, entirely surrounding the slender branches, and in groups of six to twenty or more. The reproductive spores were borne in cones or catkins at the tips of the branches in much the same manner as those of the *Lepidodendron* trees. These great tree-like rushes are represented today by a small descendant, the *Equisetum,* commonly called the horsetail, scouring-rush, or jointed grass, which grows in moist localities along streams or roadsides. The growth of secondary wood in the trunks of the ancient *Equisetales* enabled them to become sturdy tree-like forms very different from the *Equisetum* of today, which has no secondary thickening.

The fossil remains of these large rushes usually occur as pith casts of the trunks and branches in the shales or sandstones above the coal seams. Such casts are referred to the genus *Calamites* Suckow (Fig. 59). Large casts may measure a foot or more in diameter. It must be remembered that in life this pith was surrounded by a wide cylinder of secondary wood, and that this in turn was surrounded by a thick cortex.

Courtesy Chicago Natural History Museum

FIGURE 58

Restoration of *Calamites,* a gigantic rush-like plant. The foliage is classified under the name *Annularia.*

The pith casts bear longitudinal ribs and distinct joints, or nodes, which may sometimes be constricted. Frequently, the impressions have lost their cylindrical form because of crushing prior to fossilization. For a long time workers believed that these pith casts with their distinctive ribs represented the outer surface of the stems. Actually, fossils of the outer stem surfaces are rare, and when found are generally quite smooth and only occasionally very slightly ribbed. Such impressions of the outer stem surfaces may be referred either to *Calamites* proper, or to *Calamophyllites* Grand 'Eury, a

Photograph by George Langford

FIGURE 59

Impression of a *Calamites* stem showing mode of branching. (X ½)
ISM 14774

special "genus" for impressions of the outer surfaces. Arber subdivided this into two special genera: *Calamophloios,* in which the nodes between the joints are smooth or only very slightly striated longitudinally; and *Dictyocalamites,* in which the striations on the internodes are reticulate, or net-like. The latter are very rare. These special genera have not been adopted generally by paleobotanists.

Weiss divided the genus *Calamites* into three subgenera, based on the manner of branching: *Calamitina,* in which the branch-bearing nodes were separated regularly by a definite number of branchless nodes, usually three; *Eucalamites,* in which all nodes bore branches; and *Stylocalamites,* in which the branches were very few, were scattered widely along the trunk and were often separated from each other by a great number of branchless nodes. This method of classification, while good in principle, has little practical value because of the vast number of fossils which are found in fragmentary conditions such that the scheme of branching and the nodal series cannot be observed accurately. The sub-generic names are not used in the naming of species.

Similarly, as in the case of the lycopod trees, fossils occur which show various layers of the stem. Circular impressions are also found which represent transverse sections of the stems. Here again, such remains have no botanical or geological value.

In life, the pith in the trunk of *Calamites* broke down at a relatively early stage leaving diaphragms, or rings, of tissue at the nodes. The fossils, which were formed by sediments filling these pith cavities, consequently show the markings on the inner surface of these woody cylinders. These markings, which are diagnostic features in classifying the various species, may be listed as follows:

1. Nodes (the joints), and internodes (the stem areas between the joints). In some species the nodes are constricted to give a waist-like aspect.
2. Longitudinal furrows on the internodes, caused by wedges of wood projecting into the pith cavities.
3. Raised ribs between the furrows, indicating the positions of the medullary rays.

4. Small knobs, or tubercles, just above and below the node-line may be observed on well preserved specimens. They mark the position of canals of soft tissue, called the infranodal canals, which passed through the medullary rays and opened into the pith.
5. Large oval or circular scars may exist on, or slightly above, some of the nodes. These mark the attachment points of former branches.

FIGURE 60

Calamites suckowi Brongniart. (X 1)
ISM 15629

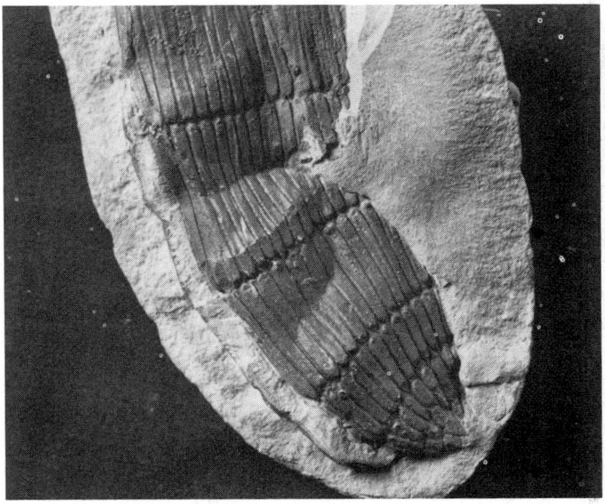

Photograph by George Langford

Calamites suckowi Brongniart (Fig. 60) is the most common Illinois species. The internodes, which vary from one to eight inches in diameter, and up to six inches in length, are characteristically wider than high. The ribs are straight, half-round or planoconvex, and are bluntly obtuse at the ends where they meet the nodes. The nodes are not constricted. The upper tubercles are relatively large and are oval or circular; lower ones are usually absent, as are also branch scars. The ribs are usually marked by fine longitudinal striations.

Calamites cisti Brongniart (Fig. 61) has internodes that are characteristically longer than wide, frequently being very slender. They may vary from three-quarters of an inch to six inches in height, and the width from a half inch to four inches. The ribs are tapered slightly toward the ends and terminate in a rather blunt, bow-shaped tip. They are longitudinally striated, but this often is not visible except under the lens. The tubercles at the upper ends of the ribs are elliptical in shape. Branch scars are usually absent,

FIGURE 61

Calamites cisti Brongniart. (X 1)
Collected near Galesburg, Illinois

Photograph by author

although a few have been recognized on shortened joints near the upper part of the stem.

Calamites ramosus Artis (Fig. 62) also has internodes that are much longer than they are wide. In marked contrast to *C. cisti,* however, every node bears branch scars. Usually there are two such scars located on opposite sides of the stem, but the number may vary from one to three on each node. The ribs are straight, flat, and the ends are rounded or bluntly terminated. Oval tubercles may occur at both ends of the ribs, but often they are indistinct or absent.

Calamites cruciatus Sternberg (Fig. 63) may be differentiated from *C. ramosus* by its internodes which, like *C. suckowi,* are characteristically wider than high. In addition, every node may contain from three to six or more branch scars.

Calamites undulatus Sternberg (Fig. 64), as its name indicates, is characterized by its undulating, or wavy ribs. The internodes usually decrease in length by equal amounts progressively from one series to the next, except for the

FIGURE 62
Calamites ramosus Artis. (X ⅜)
ISM 14311

internode immediately above the branching node. This one is usually much shorter than the others. The undulating ribs are flat or slightly protruding, the ends of the ribs forming angles of approximately ninety degrees. Tubercles may usually be seen at both ends of the ribs, those at the upper

FIGURE 63

Calamites cruciatus Sternberg. (X 1) Collected in the Saar, Germany. (From Jongmans & Kidston)

Photograph by author

FIGURE 64

Calamites undulatus Sternberg. (X 1) From a coal mine near Danville, Illinois.

FIGURE 65

Calamites goepperti Ettingshausen. (X 1) Collected at Doncaster, England. (From Jongmans & Kidston)

ends being circular and not over one millimeter in diameter, and those at the lower ends being mere dots. The branch-bearing nodes are separated by numerous branchless nodes.

Calamites goepperti Ettingshausen (Fig. 65) is characterized by nodes having branch scars set so closely together that the scars touch each other. Where these are in contact their sides are flattened, thus giving the scars a rather squared outline. The arrangement of these branch-bearing nodes is irregular, there being no definite number of branchless nodes between them.

FIGURE 66

Calamophyllites goepperti Ettingshausen is the external surface of *Calmites goepperti.* (X 1)

ISM 14312

Calamophyllites goepperti Ettingshausen (Fig. 66) represents the impressions of the outer surface of the stem of *Calamites goepperti.* Ribbing is faintly visible. The branch scars touch each other and are flattened sideways to give a squared outline like those on the pith casts.

The leaves of *Calamites* were borne in whorls completely surrounding the branches, and because of their characteristic appearance they have often been called fossil asters by the coal miners. The individual leaves of a given whorl were often of unequal length. They are classified under two "genera," *Annularia* Sternberg and *Asterophyllites* Brongniart. The leaves of *Annularia* may be linear, lanceolate, or spathulate, and are spread out in a single plane. They are united at the bases to form a small collar around the stem,

FIGURE 67

Annularia stellata Martin has the widest portion of the leaves slightly beyond the middle. (X 3/8)

ISM 14275

but this feature cannot be seen on most specimens. The leaves of *Asterophyllites* are proportionately longer and narrower than those of *Annularia,* and are neither spread out in a single plane nor united at the bases. A single vein exists in both types of leaves.

Annularia stellata Martin (Fig. 67) has long, obtusely pointed, spathulate leaves which have their widest point above the middle. There are from sixteen to thirty-two leaves per whorl, and the lateral leaves are somewhat longer than the others. The leaves of the successive whorls along the stem may touch or slightly overlap the next whorl.

Photograph by George Langford

FIGURE 68

Annularia radiata Brongniart has the widest portion of the leaves near the middle. (X ½)
ISM 14775

Annularia radiata Brongniart (Fig. 68) has long, linear, sharply pointed leaves which have their widest point near the middle. There are eighteen to twenty-four leaves per whorl, and they are all about equal in length. Sometimes the successive whorls overlap each other.

Annularia sphenophylloides Zenker (Fig. 69) has small, spathulate leaves which have their widest point above the middle. The leaves are sharply pointed and sometimes a tiny thorn-like extension may be observed on the tip. There

FIGURE 69

Annularia sphenophylloides Zenker. (X 4/5)
ISM 14272

are twelve to eighteen leaves per whorl and, as in *A. stellata,* the lateral ones are somewhat longer. The successive whorls usually touch each other.

Annularia galioides Lindley & Hutton (Fig. 70) has rather small, sharply pointed leaves which, like *A. radiata,* have their widest point in the middle. All are usually about the same length. The number of leaves per whorl is about twelve.

FIGURE 70

Annularia galioides Lindley and Hutton. (X 5/4)
ISM 14266

FIGURE 71

Asterophyllites equisetiformis Schlotheim. (X 3/5)
ISM 14291

Asterophyllites equisetiformis Schlotheim (Fig. 71) has very long, sharply pointed, needle-like leaves. The individual leaves are equal to, or somewhat longer than, the internodes

FIGURE 72
Asterophyllites longifolius Brongniart. (X ½)
ISM 14293

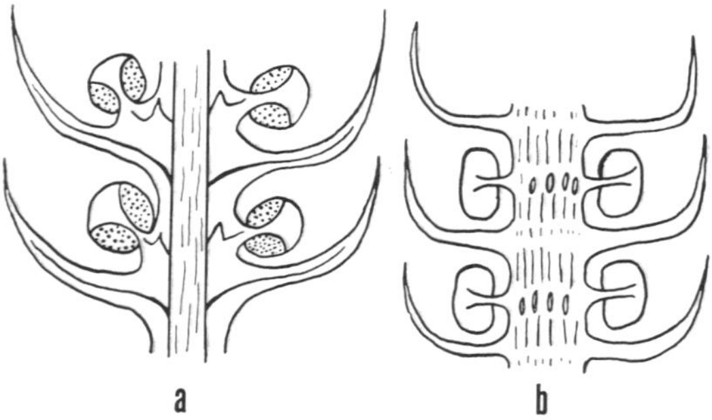

FIGURE 73
Diagram showing mode of attachment of spore cases on cones of *Calamites*. a. *Paleostachya* Weiss (after Renault). b. *Calamostachys* Schimper (after Zeiller, X 3)

between the whorls. The number of leaves in each whorl may vary from twelve to twenty.

Asterophyllites longifolius Brongniart (Fig. 72) is similar in most respects to *A. equisetiformis* except that the individual leaves are much longer, usually about twice as long.

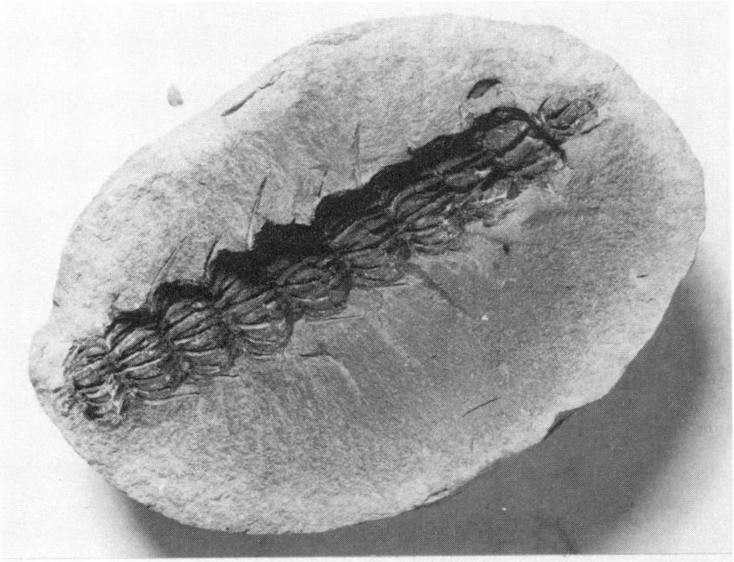

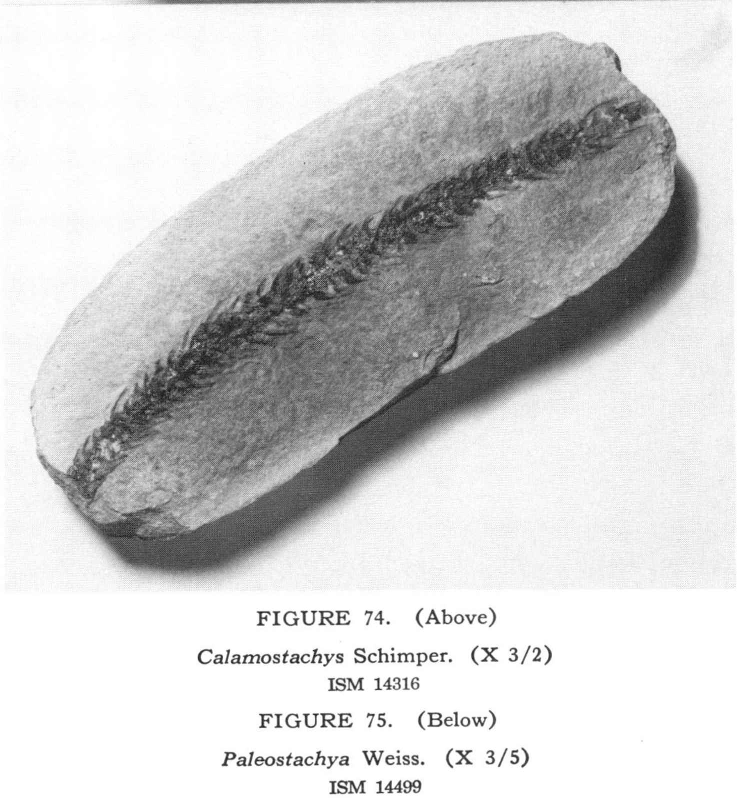

FIGURE 74. (Above)

Calamostachys Schimper. (X 3/2)

ISM 14316

FIGURE 75. (Below)

Paleostachya Weiss. (X 3/5)

ISM 14499

The fructifications of *Calamites* consisted of rather long, slender cones which were borne on the tips of the branches, much after the manner of the *Lepidodendrons*. Similarily, the cones were shed when ripe so that fertilization could occur on the swampy ground. The averaged sized specimens are from two to three inches long, by a half inch or less in width. Unusually large types may be six to twelve inches long, by an inch or an inch and a half wide. They are classified according to their structure, and in poorly or imperfectly preserved specimens it is difficult to differentiate between the several forms.

Calamostachys Schimper (Figs. 73, b & 74) has narrow cones consisting of alternate whorls of sterile bracts and spore cases attached at right angles to the stem axis. The spore cases are attached midway between the bracts. Cones of the *Calamostachys* type have been found attached to branches bearing leaves of *Annularia stellata, A. sphenophylloides,* and to *Asterophyllites*.

Paleostachya Weiss (Figs. 73, a & 75) is quite similar to *Calamostachys,* and poorly preserved specimens cannot be differentiated. In this genus the spore cases are attached in the axils of the sterile bracts instead of midway between them as in *Calamostachys*.

Macrostachya Schimper (Fig. 76) is larger than the other forms, attaining lengths up to eight inches, and widths of an inch and a quarter. The whorls of bracts are much closer together. Each whorl contains about twenty bracts, each united to the other. Near their tips, however, the bracts divide into sharply pointed teeth.

Volkmania Sternberg includes cone impressions having the general aspect of *Calamostachys* and *Paleostachya,* but are preserved in such a way that the attachment of the spore cases in relation to the sterile bracts has never been observed. As a result, they cannot be assigned to either of these species.

The roots of *Calamites* show no distinguishing features and cannot be readily classified when detached from the parent plant. Except for the tree roots, *Stigmaria,* those of all other coal age plants are quite similar. Consequently the remainder are all grouped together under the "generic" name *Pinnularia* Lindley and Hutton (Fig. 77).

FIGURE 76

Macrostachya Schimper. (X 1) Collected in Illinois; exact locality unknown.

Photograph by author

FIGURE 77

Pinnularia Lindley and Hutton represents the rootlets of *Calamites* and other coal age plants. Collected near Galesburg, Illinois. (X 1)

Photograph by author

CHAPTER V

Sphenophyllales

THE ORDER *Sphenophyllales* includes a single genus only, *Sphenophyllum* Brongniart (Fig. 78), which became extinct with the close of the coal period. The plant was herbaceous in habit and was relatively small, consisting of a slender stem bearing whorls of wedge-shaped leaves. The species are based on variations in the leaf form which ranged from nearly entire, triangular forms to deeply cleft, linear forms. In appearance, *Sphenophyllum* resembled somewhat the modern bedstraw *(Galium aparine)*, but actually the two bear no relationship.

The leaves of *Sphenophyllum* were attached around the stem in multiples of three, with six to twelve being the average number per whorl in most species. The leaf bases were separate and did not fuse to form a collar around the stem as was true of the Annularian leaves among the *Calamites.* One or more veins, varying with the species, entered the base of each leaf and forked repeatedly until a final single veinlet passed into each ultimate segment of the leaf. The more common species had leaves with broad apexes cut up into numerous teeth. These teeth were rounded in some species and pointed in others. In some cases the leaves varied in shape upon the same plant, the larger branches tending to bear the less divided ones, and the smaller branches having more deeply cleft individuals.

Sphenophyllum emarginatum Brongniart (Figs. 79, a & 80) is the most common species in Illinois. It has two types of leaves, which vary in length from five to ten millimeters,

Courtesy Chicago Natural History Museum

FIGURE 78

Restoration of *Sphenophyllum,* a small herbaceous plant forming much of the undergrowth in the coal age forests.

with widths of two and a half to seven millimeters across the apex. The smaller ones are undivided except for six small rounded teeth along the apical margin. The larger leaves are cleft in the middle, each segment bearing four to six rounded teeth. One or two veins may enter the base of the leaf of either type.

Sphenophyllum majus Bronn (Figs. 79, c & 81) is quite similar to *S. emarginatum* but is at least twice as large, the leaves being twelve to twenty millimeters in length. In addition, the apical teeth are pointed, not rounded as in the previously described species. The larger leaves frequently have three clefts, the middle one being deeper than the lateral ones. The number of teeth varies from eight to sixteen. Two veins enter the base of the leaf.

Sphenophyllum longifolium Germar (Fig. 79, e) may reach lengths of an inch or slightly more. It varies from the preceding species in that four veins enter the base of the leaf; also, the middle cleft is very deep, extending down to the middle of the leaf or beyond. There are twelve or more sharply pointed teeth.

Sphenophyllum myriophyllum Crépin (Fig. 79, d) has leaves varying in length from ten to thirty millimeters. It has three very deep clefts reaching nearly to the leaf base, which gives it a resemblance to a deer antler. The teeth are pointed, and one or two veins may enter the base.

Sphenophyllum cuneifolium Sternberg (Fig. 79, b) has two types of leaves which are about the same size as those of *S. emarginatum*. One form having six to twelve teeth bears a superficial resemblance to *S. emarginatum* except that the teeth are pointed, not rounded. The other form is similar in aspect to *S. myriophyllum* except that the clefts do not extend beyond the middle of the leaf and no more than one vein enters the base.

Sphenophyllum tenuifolium Fontaine & White (Figs. 79, f & 82) has long, linear leaves with very few sharply pointed teeth, usually about five in number. The leaves

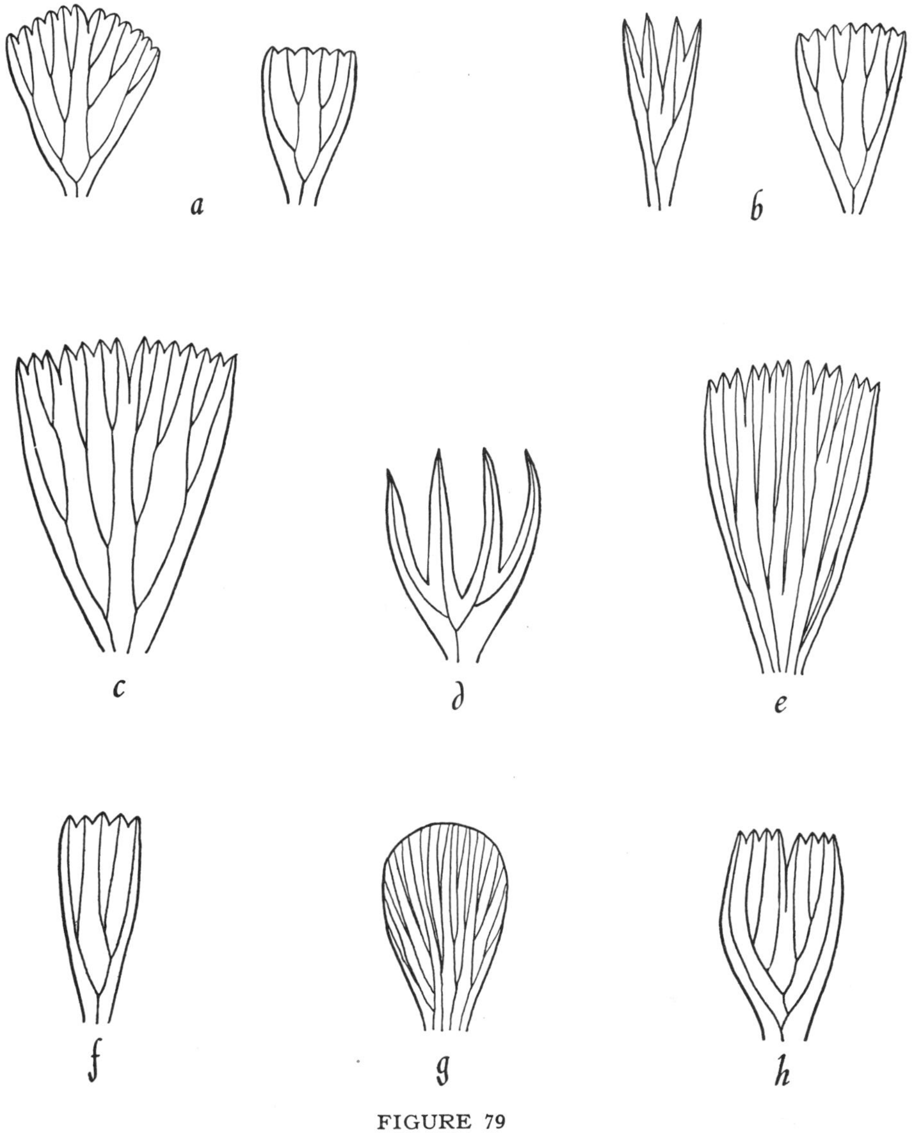

FIGURE 79

Leaf forms of *Sphenophyllum*.

a. S. *emarginatum* Brongniart, both forms (X 3);
b. S. *cuneifolium* Sternberg, both forms (X 4);
c. S. *majus* Bronn (X 3);
d. S. *myriophyllum* Crépin (X 2);
e. S. *longifolium* Germar (X 2);
f. S. *tenuifolium* Fontaine & White (X 2);
g. S. *verticillatum* Schlotheim (X 3);
h. S. *oblongifolium* Germar & Kaulfuss (X 4).

average ten millimeters in length by five or less in width. Usually one vein only enters the base.

Sphenophyllum oblongifolium Germar & Kaulfuss (Fig. 79, h) is characterized by convexly bowed leaves having

FIGURE 80

Sphenophyllum emarginatum Brongniart. (X 7/5)
ISM 14538

their widest point near the middle. The leaves are about ten millimeters in length and have a single cleft which does not extend to the middle. There are usually about four pointed teeth on each segment. Only one vein enters the base.

Sphenophyllum verticillatum Schlotheim (Fig. 79, g) differs from other forms in that the leaf margin is distinctly rounded so that the widest portion is just below the apex.

The latter may be smooth or very finely serrated. There are always more than two veins, usually four, entering the base of the leaf.

The fructifications of *Sphenophyllum* are cones similar in general aspect to those of the *Equisetales*. They reached

Courtesy Illinois State Geological Survey

FIGURE 81

Sphenophyllum majus Bronn. (X 1)

lengths of nearly five inches, and widths of a half inch. The cones bore numerous bracts whose bases were fused to form a collar around the cone axes. The spores occurred in cases which were borne directly on the upper surfaces of some or all of the bracts, depending on the species. Detached specimens are referred to the genus *Sphenophyllostachys* Seward, but when found attached they are given the name of the associated leaf species. Except that they are usually longer in relation to their width than the cones of the *Calamites,* specimens which are not sufficiently well preserved to show details of their structure cannot be readily differentiated.

FIGURE 82

Sphenophyllum tenuifolium Fontaine and White. (X 1)

ISM 14776

Photograph by George Langford

CHAPTER VI

Filicales and Cycadofilicales

IMPRESSIONS of fern-leaf foliage in the shales of the coal age are scarcely distinguishable in appearance from the finely divided fronds of present-day ferns. Until quite recently these plants were all regarded as true ferns *(Filicales)*, and because of their abundance the coal period was

Courtesy Chicago Natural History Museum

FIGURE 83

Restoration of a seed-fern, or *Cycadofilicales*. The seeds were borne at the tips of the pinnae.

often called the Age of Ferns. Modern research, however, has shown that the stem structure of many of these so-called ferns differed from that of any living fern, and more recently, fossils have been discovered showing seeds attached to the "fern" fronds. This is in marked contrast to all modern ferns which reproduce by means of spores borne on the under surfaces of the leaflets. In these respects the fossil plants more closely resembled the modern cycads, sometimes erroneously called cycad-palms because of their linear type of foliage. These fossil plants have therefore been placed in a separate order, called *Cycadofilicales* (cycad-like ferns) by American paleobotanists, and *Pteridospermeae* (seed-ferns), by European paleobotanists (Fig. 83). The popular term, seed-ferns, is also used in America. Some of the coal age plants were true ferns, however, but they occupied a subordinate position to the seed-ferns and cannot be distinguished from them except when spore cases are found on the fertile leaflets.

Both ferns and seed-ferns are characterized by relatively large leaves, called fronds, which are sub-divided numerous times into several orders of pinnae. The pinnae are divided into ultimate units called pinnules. Because of these numerous units, the fronds as a whole present a delicate lacey appearance. The reproductive organs are borne directly on the fronds, the true ferns having their spore-cases attached to the under sides of the pinnules, and the seed-ferns having borne nut-like seeds at the tips of the stems, or rachises, of the leaves. Both groups included herbaceous forms with habits similar to modern ferns. Other forms were scramblers or vines, and still others were trees reaching heights of seventy feet with stem diameters up to two feet or more. The tree-ferns, having stems clothed in a thick mass of descending roots, and bearing a crown of massive spreading fronds, were not vastly different from those of our present-day tropics, except that some forms were larger and some bore seeds. The seed-ferns became extinct shortly after the close of the coal forming period, but true ferns have continued to live through all of the succeeding changes of the earth's geologic history down to the present time.

Many species of fern-like leaves occurring as fossils cannot, as yet, be placed in their correct botanical positions inasmuch as reproductive organs or stems with diagnostic features have not been found in association with them. Hence we do not know whether such species were ferns or seed-ferns. Consequently it is more practical to consider all

FIGURE 84

Restorations of two forms of tree-ferns. Some of these massive ferns reached heights of seventy feet during the coal age.

Courtesy Chicago Natural History Museum

of the fern-like leaves as a unit. They are classified into "form" genera and species, based on the shape, venation, and mode of attachment of the pinnules. When leaves provisionally classified in this manner are later found with reproductive organs attached, they are removed to other appropriate genera. Most of the fern-like fossils found in Illinois are classified in the following form genera.

Sphenopteris Brongniart (Figs. 90 to 92) has leaves that are very lacey in general appearance. The ultimate divisions of the leaves, or pinnules, are quite small and greatly lobed or dissected; they are usually contracted at the bases. The veins radiate fan-like from the base and usually fork several times so that at least one veinlet is continued into each segment of the pinnule. Some Sphenopterids are ferns, some are seed-ferns, and the affinities of others are not known.

Diplothmema Stur (Figs. 94 & 99, c) closely resembles some forms of *Sphenopteris,* but is distinguished by its more deeply dissected pinnules which give it a sort of palmate appearance. It is a seed-fern.

Mariopteris Zeiller (Figs. 100 to 102) has pinnules which frequently present a thick, leathery appearance and have broad bases. They are generally less segmented than those of *Sphenopteris,* and the venation may be coarser and more conspicuous. It is a seed-fern.

Pecopteris Brongniart (Figs. 103 to 125) is characterized by its pinnules which are not contracted at the bases and are attached to the stem, or rachis, by their entire breadth. Each pinnule has a distinct midrib which continues from base to apex, or within a very short distance of the apex. Numerous lateral veinlets are given off from the midrib at almost right angles to it, and these in turn may be simple or may fork one or more times before reaching the margin of the pinnule. The tips of the pinnules in most species are rounded and the sides are parallel or nearly so. Other species present a saw-tooth aspect, the tips of the pinnules being pointed and the sides distinctly not parallel. *Pecopteris* is one of the most abundant forms of the coal age and includes very many species. Some are ferns and others may be seed-ferns. The affinities of many are still unknown.

Aphlebia Presl (Fig. 98) is a leaf form which occurred near the bases of fronds belonging to various Sphenopterid and Pecopterid plants. Many of these leaves bear a vague resemblance to those of ordinary garden lettuce. Some forms are very deeply cleft and segmented. The venation is forked or indistinct. Specimens are usually found isolated and consequently cannot be referred to their parent plants. There is a wide variety of forms and nearly every specimen differs somewhat from the next. Several species are recognized, but they are of little value. Such leaves are usually referred to the genus without attempt being made to classify them specifically.

Neuropteris Brongniart (Figs. 137 to 147) has pinnules that are oval or tongue-shaped with cordate, or heart-shaped, bases. They are attached to the rachis by a single point at the base of the midrib. The midrib does not continue all the way to the apex, however. Entering the pinnule at the point of attachment, shortly thereafter it breaks up into numerous secondary veins which arch toward the borders. Each secondary vein may fork one to many times, the numerous tiny veinlets so formed arching more or less parallel to each other. The forking veinlets never come together again as they approach the margin, nor do they come into contact with veinlets from adjacent veins. *Neuropteris* also was one of the most abundant forms of the coal age. It is a seed-fern.

Desmopteris Stur (Figs. 135 & 136) has pinnules that are much larger and longer than those of *Alethopteris,* and are not decurrent. The bases are broad and are attached directly to the rachis by their entire width. There is a distinct midrib continuous to the apex; the lateral veinlets fork once in some species, and as many as four times in others. It is a seed-fern.

Alethopteris Sternberg (Figs. 126 to 134) has pinnules which more nearly resemble those of *Pecopteris* than any other form. Certain species of each seem to grade into one another and it is frequently difficult to draw the line between them. The pinnules of *Alethopteris* are characteristically longer and more linear in general aspect, however. Usually they are decurrent, that is, the bases of adjoining pinnules coalesce into one another at their points of attachment along the rachis. The midrib is continuous to the apex of the pinnule. The numerous lateral veinlets are simple, or forked only once or twice. It is a seed-fern.

Cyclopteris Brongniart (Figs. 148 & 149) includes orbicular and kidney-shaped leaves, with radiating venation, which occurred at the bases of composite leaves and fronds of *Neuropteris*. These leaves are usually found isolated, and for this reason were originally thought to be a distinct genus. They are now known to be the specialized leaves of various Neuropterids just as *Aphlebia* are specialized leaves in the Sphenopterid group. Again, only when found attached, can we be sure as to the exact species to which a specimen may belong. It is therefore convenient to retain the name *Cyclopteris* for the inclusion of detached specimens.

Odontopteris Brongniart (Figs. 154 to 156) has pinnules of much the same general shape as in *Neuropteris,* but they are attached by the whole breadth of their bases and are often decurrent. There is no distinct midrib; several veins enter the pinnule from the rachis, then fork several times, producing secondary veins which arch to the margins. It is a seed-fern.

Mixoneura Weiss (Figs. 150 & 151) occupies a mid-position between *Neuropteris* and *Odontopteris*. The shape of the pinnules is the same as in *Neuropteris,* and they are attached to the rachis by a single point. The midrib, however, is represented by several strands which fork numerous times to produce the arching veinlets. Before this genus was established by Weiss all of the species were included in *Neuropteris*. It is a seed-fern.

Linopteris Presl (Figs. 152 & 153) also has pinnules of the same shape as those of *Neuropteris*. The venation, however, is quite distinctive. There is no definite midrib, and the veins anastomose, or reticulate, to form a network over the entire surface of the pinnules. It is a seed-fern.

Spiropteris Schimper (Fig. 85) includes young leaves or tips of older ones which were fossilized while still in the coiled condition. Because of their immaturity it is not possible to recognize the pinnule form when detached from the parent plants. It is convenient, therefore, to refer such specimens to this genus without attempting specific classification. They are, of course, members of both orders, the *Filicales* and *Cycadofilicales*.

Fossil impressions of the stems, or trunks, of tree-ferns are often found. They bear numerous large scars representing the former points of attachment of the fronds. Within the large scar is usually seen an oval or horse-shoe shaped scar marking the point where the vascular bundle passed from the stem to the rachis of the frond. These fossil stems are referred to three form genera.

FIGURE 85

The coiled tips of fern fronds are classified under the name *Spiropteris* Schimper. (X 1) ISM 14553

Caulopteris Lindley & Hutton (Fig. 86) includes stems having oval or circular scars which are spirally arranged entirely around it.

Megaphyton Artis (Figs. 87 & 88) has its scars arranged in two vertical rows on opposite sides of the stem, those on one side alternating in position with those on the other.

Courtesy Illinois State Geological Survey

FIGURE 86

Caulopteris Lindley and Hutton represents the stems of tree-ferns having branch scars spirally arranged around the stem. (X ⅓)

Ptychopteris Corda (Fig. 89) includes decorticated, or sub-surface, layers of both of the above two types. As in the case with the lycopod and gymnospermous trees, the impressions of these underlayers have little practical value.

FIGURE 87

Impressions of a single scar of a tree-fern belonging to the species *Megaphyton insignis* Lesquereux. (X ⅔)
ISM 14555

Photograph by George Langford

Detached roots of ferns and seed-ferns cannot be differentiated. They are included in the form genus *Pinnularia* Lindley & Hutton (Fig. 77), along with the roots of other coal age plants.

FIGURE 88

Megaphyton Artis represents the stems of tree-ferns having branch scars arranged in vertical rows. (X ⅓)

Courtesy Illinois State Geological Survey

Returning again to the consideration of the foliar structures of the ferns and seed-ferns, it must be remembered that the genera and species established by the early paleobotanists were based almost entirely upon the form of the leaves. In many cases these original classifications still hold today. As more fossils are unearthed and knowledge regarding them increases, it is frequently necessary to impose restrictions on the use of original names, or to institute new ones for the reception of specimens which do not conform to the original descriptions. In this way many leaves originally included in the genus *Pecopteris,* for example, are now placed in the newer genera *Asterotheca, Dactylotheca, Ptychocarpus,* etc. Thus *Pecopteris miltoni* Artis is now known as *Asterotheca*

FIGURE 89

Ptychopteris Corda represents the sub-surface bark of tree ferns. (X ⅓)
ISM 14752

miltoni Artis. Such changes have resulted from the discoveries of newer specimens containing reproductive organs attached to the leaves, similar types of which were previously known only in the sterile condition. Since the reproductive organs on many of these Pecopterid leaves vary in structure, they could not all be retained within a single genus, and the necessity of setting up new genera is obvious. Conversely, as previously mentioned, the opposite situation is sometimes true, as in the case of *Cyclopteris* which belongs to various species of *Neuropteris*.

The basis upon which revisions in classification are made is often dependent upon differences in structure that cannot be observed except under the microscope. Even then, it is often only the unusually well preserved specimen that may show them. The systematic position of the vast majority of fern-like impressions is known only by comparison with such individual specimens that have been meticulously studied. For example, *Asterotheca* is a fern which produced spores in cases called sporangia, of which four or five were united in a cluster called a synangium. Many of these synangia were borne on a single pinnule. *Ptychocarpus,* on the other hand, had five to eight sporangia united to form a synangium. Such microscopic differences as these, which cannot be observed on the average specimen, are beyond the scope of this handbook and will not be discussed in detail.

The various species of fern-like impressions commonly occurring in Illinois have been grouped according to the form of their pinnules, and for all practical purposes this scheme will be adequate for ready identification. Where reclassifications based on structure have occurred, however, the revised names are used. Some groups, therefore, will include species whose generic distinction differs in name from the "form" name of the group.

Sphenopterid Group

Sphenopteris obtusiloba Brongniart (Figs. 90 & 99, j), as its name indicates, is characterized by its obtusely lobate pinnules with bases that are contracted into a narrow footstalk. The central vein gives rise to many close, fine, arching, secondary veinlets, some of which fork several times before reaching the margins.

Photographs by author

FIGURE 90
Sphenopteris obtusiloba Brongniart. (X 1) Collected near Galesburg, Illinois.

FIGURE 91
Sphenopteris artemisaefolioides Crépin. (X 1) Collected near Galesburg, Illinois.

Sphenopteris artemisaefolioides Crépin (Figs. 91 & 99, f) has pinnules that present a rather elongate-ovate appearance. A single vein enters from the rachis, forking several times so that one or two veinlets enter each obtusely rounded segment of the pinnule.

Sphenopteris spiniformis Kidston (Figs. 92 & 99, i) is very similar in general aspect to *S. artemisaefolioides*. In this species, however, the apexes of the pinnules are sharply pointed.

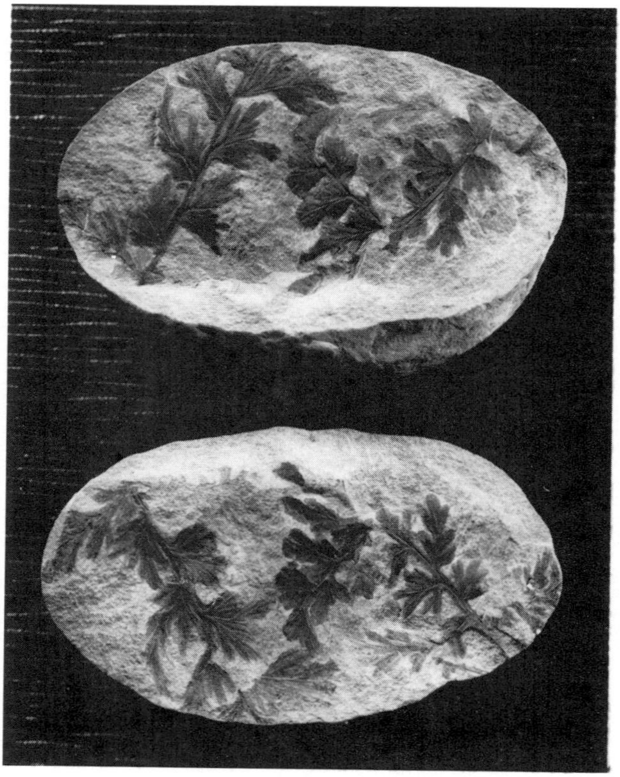

Photograph by George Langford

FIGURE 92

Sphenopteris spiniformis Kidston. (X 1)
ISM 14550

FIGURE 93

Eremopteris missouriensis Lesquereux. (X 5/3)
Collected near Colchester, Illinois.
ISM 1630

Eremopteris missouriensis Lesquereux (Figs. 93 & 99, k) has pinnules quite similar to those of *S. spiniformis*. However the pinna, or leaflet comprising the individual pinnules, is more elongate and has a more extended tip. The genus occupies a mid-position between *Sphenopteris* and *Diplothmema*.

Diplothmema furcatum Brongniart (Figs. 94 & 99, c) has pinnules which are deltoid or rhomboidal in general form and are very deeply cleft. The midrib forks several times,

FIGURE 94

Diplothmema furcatum Brongniart. (X 1)
ISM 14777

Photograph by George Langford

one veinlet extending to each segment and terminating at the sharp tip. The deeply cleft segments give the pinnules a palmate appearance.

Renaultia gracilis Brongniart (Figs. 95 & 99, g) has pinnules which are emarginate, or notched, with sharply pointed

FIGURE 95

Renaultia gracilis Brongniart. (X 1)

Courtesy Illinois State Geological Survey

tips. The midrib is slightly undulating, or flexuous, and gives rise to secondary veinlets which may be simple or forked once again, depending on the size of the pinnule. One veinlet extends to each sharp tip of the pinnule lobes. The bases of the pinnules are broad.

Renaultia chaerophylloides Brongniart (Figs. 96 & 99, h) bears a close resemblance to *R. gracilis* except that the tips of the pinnule segments are rounded instead of being sharply pointed.

FIGURE 96

Renaultia chaerophylloides Brongniart. (X 1)

Photograph by author

Corynepteris sternbergi Ettingshausen (Figs. 97 & 99, e) has small quadrate, acute, decurrent pinnules arranged along the rachis in a distinct saw-tooth manner. Each pinnule has three or four pointed teeth with one veinlet extending to each point. The sterile pinnules of this species were formerly classified under the genus *Alloiopteris* Potonié.

FIGURE 97

Corynepteris sternbergi Ettingshausen. (X 2)
ISM 14265

Aphlebia spinosa Lesquereux (Fig. 98) has a broad, flat, membranous rachis along which the pinnules are arranged. Each pinnule has several short spine-like teeth and an elongated terminal lobe. Simple veins extend into each segment.

FIGURE 98

Aphlebia spinosa Lesquereux. (X 7/10)
ISM 14277

FIGURE 99

Pinnules and pinnae of Sphenopterid and Mariopterid Groups

a. *Mariopteris nervosa* Brongniart (X 2);
b. *Mariopteris decipiens* Lesquereux (X 2);
c. *Diplothmema furcatum* Brongniart (X 2);
d. *Mariopteris muricata* Schlotheim (X 2);
e. *Corynepteris sternbergi* Ettingshausen (X 7);
f. *Sphenopteris artemisaefolioides* Crépin (X 1);
g. *Renaultia gracilis* Brongniart (X 4);
h. *Renaultia chaerophylloides* Brongniart (X 6);
i. *Sphenopteris spiniformis* Kidston (X 4);
j. *Sphenopteris obtusiloba* Brongniart (X 1);
k. *Eremopteris missouriensis* Lesquereux (X 4).

a
b
c
d
e
f
g
h
i
j
k

FIGURE 100

Mariopteris nervosa **Brongniart.** (X 1)

Photograph by
George Langford

FIGURE 101

Mariopteris muricata Schlotheim. (X ¾)
ISM 14591

FIGURE 102

Mariopteris decipiens Lesquereux. (X 9/10)

ISM 14425

Mariopterid Group

Mariopteris nervosa Brongniart (Figs. 99, a & 100) has rather large, leathery pinnules, triangular in general form, more or less united at the bases and attached to the rachis by their entire breadth. The margins of the pinnules are smooth and undissected. A single, distinct midrib enters the pinnule, giving rise to heavy, lateral veins which usually fork twice.

Mariopteris muricata Schlotheim (Figs. 99, d & 101) differs from *M. nervosa* primarily by the dentate margins of its pinnules. The venation is similar, but less pronounced, and the veins in the lowermost segment of the pinnule may originate directly from the rachis instead of branching off from the midrib.

Mariopteris decipiens Lesquereux (Figs. 99, b & 102) has pinnules which may be described as being half-round, or obovate. They are attached by a broad base and are slightly contracted just above the bases. The veinlets branch off from the midrib and usually fork twice.

Pecopterid Group

Pecopteris squamosa Lesquereux (Figs. 103 & 117, c) is characterized by its thick rachis, heavily covered with spirally arranged, overlapping scales. The pinnae, or leaflets, arising from this rachis are long and slim, and are made up of numerous very small pinnules. These tiny pinnules show only a midrib in most specimens, but rarely very small, straight, simple, lateral veinlets may be seen.

Pecopteris pseudovestita White (Figs. 104 & 117, a) has relatively large oblong pinnules with bluntly rounded apexes. The sides are not quite parallel to each other, the widest portion being at the base. The lateral veinlets fork several times and arch toward the margins, meeting them approximately at right angles. In so doing, both forks, or only the upper fork of a bifurcating veinlet may divide again.

Courtesy Illinois State Geological Survey

FIGURE 103
Pecopteris squamosa Lesquereux. (X 1)

FIGURE 104

Pecopteris pseudovestita White. (X ⅔)

ISM 14600

FIGURE 105

Pecopteris cisti Brongniart. (X 1)

ISM 14507

Pecopteris cisti Brongniart (Figs. 105 & 117, b) has pinnules that are broadly ovate and only slightly longer than they are broad. The veinlets originate from a thick midrib, are rather distant, and fork once near the base after which each fork divides again in arching toward the margin.

FIGURE 106

Pecopteris obliqua Brongniart. (X 3/2)

ISM 14515

Pecopteris obliqua Brongniart (Figs. 106 & 117, e) has large pinnules that are elongate-ovate, being considerably longer than broad, with pointed tips. They flare perceptibly near the base and are attached to the rachis quite obliquely, as the name indicates. The veinlets originate from the heavy midrib, except on the lower side near the base where several may fork off directly from the rachis. They fork once or twice, depending on their length and position on the pinnule.

Pecopteris serpillifolia Lesquereux (Figs. 107 & 117, h) has leaflets, or pinnae, which are only slightly dissected into pinnules, giving the whole an escalloped aspect about the margins. One vein, branching from the rachis, enters each segment, or pinnule, and this gives rise to several simple veinlets.

FIGURE 107

Pecopteris serpillifolia Lesquereux. (X ⅔)

ISM 14516

Pecopteris clintoni Lesquereux (Figs. 108 & 117, d) has large, linear, obtusely rounded pinnules. The margins are usually smooth, but farther down the frond they may become crenulate. The veinlets are forked once or twice, and are often obscured by a villose, or hairy, surface.

Asterotheca miltoni Artis (Figs. 109 to 111 & 117, g) is the most common of fossil Pecopterid leaves in Illinois and elsewhere. It is extremely variable and the pinnules may vary in size and appearance according to the position which they occupy on the frond. Those near the tips of the individual leaflets, or pinnae, are quite small, increasing in size farther down the rachis. As the pinnules increase in size they

FIGURE 108

Pecopteris clintoni Lesquereux. (X 1)

ISM 14505

may take on a crenulated aspect. Still farther down the stem these crenulations may be of sufficient depth to give rise to distinct units, or pinnules, in themselves. This is called a pinnatifid condition, and the leaflets of *A. miltoni* are often described as being pinnately compound. The finding of small fragments of such compound leaves is often quite confusing,

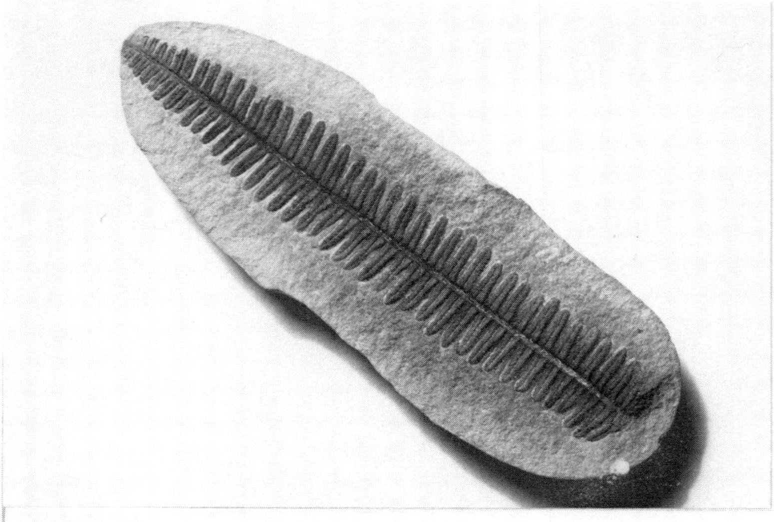

FIGURE 109. (Above)

Asterotheca miltoni Artis. (X ¾)

ISM 14305

FIGURE 110. (Below)

Asterotheca miltoni Artis. (X ½)

ISM 14307

and early paleobotanists considered as distinct species the various pinnatifid stages which are now recognized as belonging to the same plant. For example, *Pecopteris vestita* Lesquereux, a very common form, is now recognized as a pinnatifid form of *Asterotheca miltoni*. Sometimes the pinnae themselves may fork near their tips into two smaller units (Fig. 111). The individual pinnules are oblong, with parallel or nearly parallel sides, and the apexes are bluntly rounded. The lateral veinlets all arch toward the margin. The venation varies somewhat according to the size of the pinnule, but

FIGURE 111

Asterotheca miltoni Artis with forked tip. (X 1) Collected by Illinois State Geological Survey.
ISM 1760

in general it may be said that the lateral veinlets arising from the midrib fork once, after which only the upper fork divides again, the lower one remaining simple. Sometimes in the very large pinnules both forks may divide a second time. The surface of the pinnules is frequently villose, or hairy, which may obscure the veins.

Asterotheca oreopteridia Schlotheim (Figs. 112 & 117, i) has pinnules with the same general characteristics as

A. miltoni, except that the lateral veinlets fork once only, there being no further bifurcation as they arch toward the margins. The surfaces are sometimes hairy.

Asterotheca arborescens Schlotheim (Figs. 113 & 117, n) is characterized by very long, narrow pinnae which taper very gently. Sometimes irregularities in the length of the individual pinnules making up a pinna cause an unevenness in

FIGURE 112

Asterotheca oreopteridia Schlotheim. (X ¾)
ISM 1692

its tapering. The pinnules are oblong, have rounded tips, and the sides are not quite parallel, being slightly wider at the bases. The lateral veinlets may be simple or forked once only.

Asterotheca hemitelioides Brongniart (Figs. 114 & 117, f) has long, linear pinnules with parallel sides and a blunt apex. The bases are sometimes slightly contracted. The lateral veinlets are straight and simple.

Asterotheca candolleana Brongniart (Figs. 115 & 117, j) differs from *A. hemitelioides* in having arched veinlets which fork once. The bases of the pinnules are sometimes contracted and sometimes not.

Asterotheca crenulata Brongniart (Figs. 116 & 117, s), as its name implies, usually has a crenulated margin on the pinnules; however, this feature is not always observable.

FIGURE 113

Asterotheca arborescens Schlotheim. (X ½)

ISM 14296

Photograph by George Langford

The venation is quite distinctive, however. A thick midrib gives rise to thin lateral veinlets which make with it an angle of about forty-five degrees, and these generally divide again into two straight arms at a similar angle. A few of the uppermost veinlets near the apex may be simple.

Asterotheca abbreviata Brongniart (Figs. 117, t & 118) has short, blunt pinnules characterized by the presence of a midrib which is distinctly confluent, that is, it swings down along side of the rachis for a distance before uniting with it at a very sharp angle. The lateral veinlets usually fork twice, and arch to the margins of the pinnule, meeting them at an oblique angle.

FIGURE 114

Asterotheca hemitelioides Brongniart. (X 1)
ISM 14303

Acitheca polymorpha Brongniart (Fig. 117, r) is very similar to *A. abbreviata*. The midrib, however, is not confluent, but meets the rachis at nearly a right angle. The

FIGURE 115. (Above)

Asterotheca candolleana Brongniart. (X 1)

ISM 14299

FIGURE 116. (Below)

Asterotheca crenulata Brongniart. (X 1)

ISM 14512

lateral veinlets also meet the margins of the pinnule at approximately a ninety degree angle. Both forms may be quite variable, depending upon the position of the pinnules on the fronds.

Dactylotheca sturi Sterzel (Figs. 117, l & 119) has rather short pinnules with bowed sides and rather sharply rounded apexes. They are widest near the middle, and slightly contracted at the bases, giving them a somewhat ovate appearance. The lateral veinlets bifurcate once, after which the upper fork divides again, the lower one remaining simple.

Dactylotheca plumosa Artis (Figs. 117, q & 120) may be differentiated from *D. sturi* by its triangularly shaped pinnules which have a pointed apex with the widest portion at the base. The margins of the pinnules sometimes bear convex teeth, a characteristic that influenced the choice of the name given to its sterile pinnules, *Pecopteris dentata* Brongniart. This designation is still used by some authors to describe the toothed form.

FIGURE 117

Pinnule Forms and Venation of Pecopterid Group.

a. *Pecopteris pseudovestita* White (X 3);
b. *Pecopteris cisti* Brongniart (X 3/2);
c. *Pecopteris squamosa* Lesquereux (X 2);
d. *Pecopteris clintoni* Lesquereux (X 1);
e. *Pecopteris obliqua* Brongniart (X 3/2);
f. *Asterotheca hemitelioides* Brongniart (X 3/2);
g. *Asterotheca miltoni* Artis (X 4);
h. *Pecopteris serpillifolia* Lesquereux (X 1);
i. *Asterotheca oreopteridia* Schlotheim (X 4);
j. *Asterotheca candolleana* Brongniart (X 2);
k. *Ptychocarpus unitus* Brongniart, showing both forms of venation (X 2);
l. *Dactylotheca sturi* Sterzel (X 4);
m. *Crossotheca sagittata* Lesquereux, showing sterile and fertile pinnules on same frond (X 1);
n. *Asterotheca arborescens* Schlotheim, showing simple and forked types of venation (X 3);
o. *Dicksonites pluckeneti* Schlotheim (X 2);
p. *Senftenbergia pennaeformis* Brongniart, showing forked and simple types of venation (X 2);
q. *Dactylotheca plumosa* Artis (X 3);
r. *Acitheca polymorpha* Brongniart (X 3);
s. *Asterotheca crenulata* Brongniart (X 1);
t. *Asterotheca abbreviata* Brongniart (X 4).

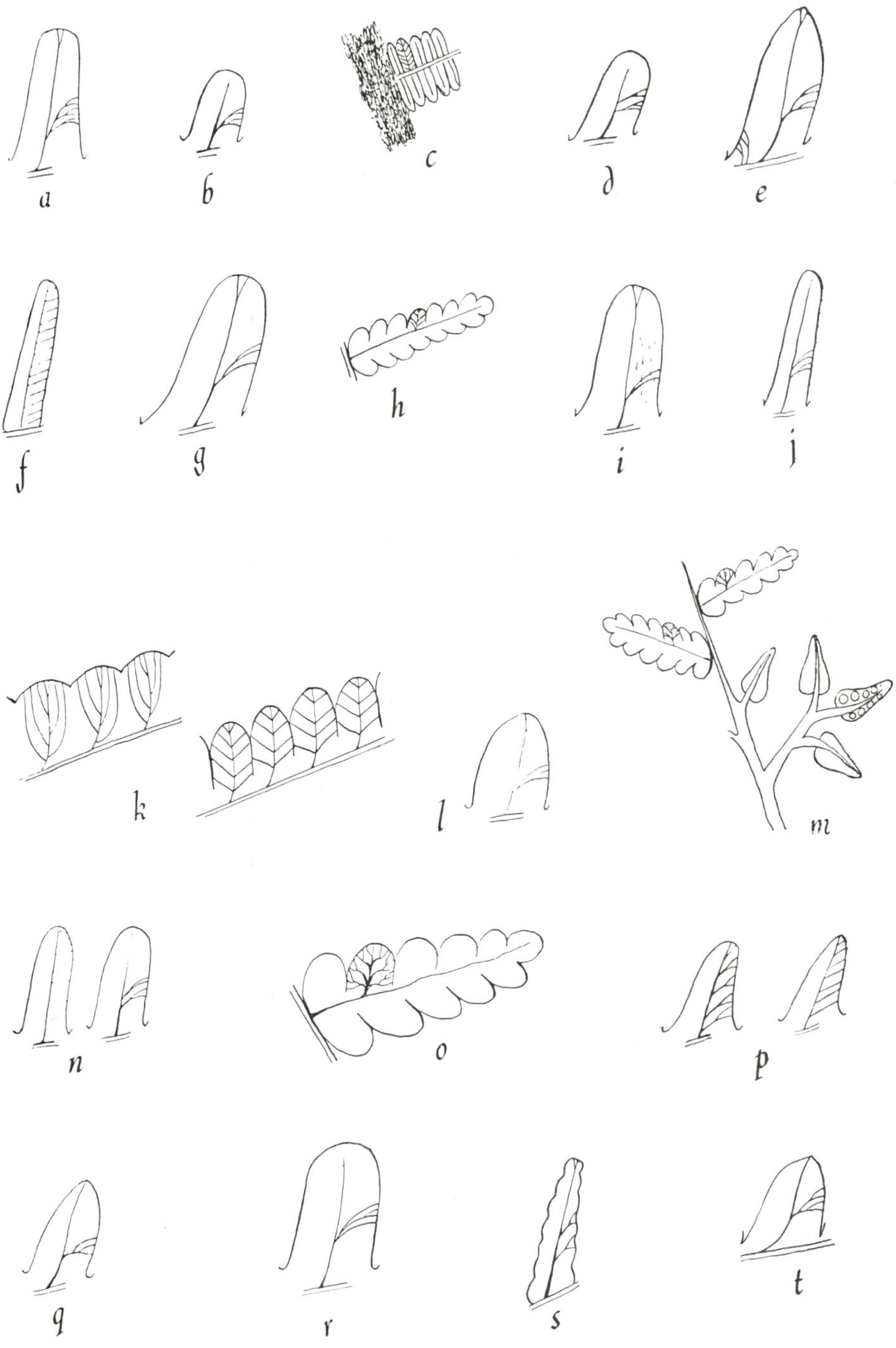
a
b
c
d
e
f
g
h
i
j
k
l
m
n
o
p
q
r
s
t

FIGURE 118

Asterotheca abbreviata Brongniart. (X 4/5)
ISM 14500

FIGURE 120

Dactylotheca plumosa Artis. (X 2/5) Collected in Illinois; exact locality unknown.
ISM 416533

FIGURE 119

Dactylotheca sturi Sterzel. (X 1)
ISM 14503

Courtesy Illinois State Geological Survey

FIGURE 121

Senftenbergia pennaeformis Brongniart. (X 1)

Senftenbergia pennaeformis Brongniart (Figs. 117, p & 121) has pinnules that are somewhat longer than those of *D. plumosa*. The venation differs in that the lateral veins are all simple or they fork once only.

Ptychocarpus unitus Brongniart (Figs. 117, k & 122) has pinnules which are situated at right angles to the rachis and are distinctly united to each other. Sometimes this union

Photograph by George Langford

FIGURE 122

Ptychocarpus unitus Brongniart showing a form of the species with veinlets emerging from the midrib directly opposite each other. (X 7/8)
ISM 14778

FIGURE 123

Dicksonites pluckeneti Schlotheim. (X 1)
ISM 14358

extends half way up the length of the pinnules with only the upper parts being free. The shape and venation of the pinnules varies according to their position on the frond. Some have the lateral veinlets emerging from the midrib directly opposite each other and extending straight to the border; in others they alternate and arch concavely to the border. In both forms the veinlets are simple.

FIGURE 124

Crossotheca sagittata Lesquereux. Specimen with upper pinnules sterile, and lower ones fertile. (X ¾)

ISM 14548

Dicksonites pluckeneti Schlotheim (Figs. 117, o & 123) is characterized by pinnules nearly as broad as they are long. They are distinctly rounded at the apex and sometimes constricted at the base, giving them a very rounded aspect as a whole. The midrib is slightly flexuous, or undulating. Some veinlets are simple, some fork twice, and some may have only one fork of a bifurcating veinlet dividing again.

Crossotheca sagittata Lesquereux (Figs. 117, m & 124-125) has large fertile pinnules which are expanded at the base in the form of an arrowhead. The upper surface is flat with a median rib, and the lateral veinlets are obscured. The spore cases, or sporangia, are unusually large, measuring from two and a half to four millimeters in length, and from one-half to three-quarters millimeters wide. They are situated in a single row entirely around the border of the pinnule. The sterile

FIGURE 125

Crossotheca sagittata Lesquereux. Specimen with all leaves fertile. (X 1)

Photograph by George Langford

pinnules are very different in appearance—so much so that had they not been found in connection, they would not have been recognized as belonging to the same plant. They are small and rounded, with lateral veinlets that fork one to three times according to the size of the pinnule. The sterile pinnules occur at the extreme apex of the frond, the fertile ones being some distance below. Sometimes all the pinnules are fertile.

Alethopterid Group

Alethopteris serli Brongniart (Figs. 126 to 128) has pinnules which vary in shape according to their position on the frond. Those at the tips of the fronds are quite short and their lower portions are strongly decurrent along the rachis. They increase rapidly in size down the rachis, becoming five or six times longer than broad. Below this the pinnatifid condition prevails in which the pinnae have a relatively large terminal pinnule with numerous very short lateral pinnules. Farther down the frond there are very large pinnae with pinnules which are more characteristic of the species. These vary in length from a half inch to two inches, and in width from three-eighths to a half inch. These linear pinnules have an obtusely pointed apex and have their widest point near the middle. Some veinlets are simple, others fork once, and a few twice.

FIGURE 126

Alethopteris serli Brongniart. (X ¾)

ISM 14253

Alethopteris lonchitica Schlotheim (Fig. 129) varies from *A. serli* in having pinnules which are longer in relation to their width, and more sharply pointed. Their size, as a whole, however, is usually smaller, averaging about an inch in length by less than a quarter of an inch in width. They are also slightly wider in the middle. The veinlets may be simple, or forked once or twice.

FIGURE 127

Alethopteris serli Brongniart. (X 2/5)
ISM 14252

FIGURE 128

Alethopteris serli Brongniart. (X 5/8)
ISM 14257

Alethopteris decurrens Artis (Fig. 130) has pinnules which are decidedly linear, with nearly parallel sides, and with the lower portion strongly decurrent along the rachis. They average about an inch in length, but do not exceed an eighth of an inch in breadth. The veinlets are simple, or fork only once.

FIGURE 129

Alethopteris lonchitica Schlotheim. (X 7/10)
ISM 14250

Photograph by author

FIGURE 130

Alethopteris decurrens Artis. (X 1) Collected in Illinois; exact locality unknown.

Alethopteris grandini Brongniart (Fig. 131) is somewhat similar to *A. serli,* but the pinnules are more rounded at the apex and less swollen at the middle. The veinlets are farther apart and usually divide twice.

Alethopteris davreuxi Brongniart (Fig. 132) can be recognized by its flexuous, or undulating, veinlets which fork once or twice. The bases of the pinnules are not contracted. They vary in shape from short, rounded forms to linear forms, depending upon their position on the frond.

FIGURE 131

Alethopteris grandini Brongniart. (X 1)
ISM 14248

Alethopteris ambigua Lesquereux (Fig. 133) is characterized by the presence of a shallow gutter entirely around the margins of the linear pinnules. On freshly opened concretions this gutter is frequently filled with a white calcite deposit which causes it to stand out prominently. Unfortunately, sometimes other species of *Alethopteris* may exhibit this feature, particularly *A. grandini*. However, the pinnules of *A. ambigua* are slightly more linear than are those of the other species. The midrib is very thick, and the veinlets fork once or twice.

FIGURE 132

Alethopteris davreuxi Brongniart. (X 9/10)

ISM 14244

Photograph by George Langford

Alethopteris sullivanti Lesquereux (Fig. 134) has pinnules which are swollen above the middle and have very round apexes. The smaller pinnules may be about equal in length and breadth, and the larger ones are about twice as long as they are broad. Some veinlets are simple, others fork

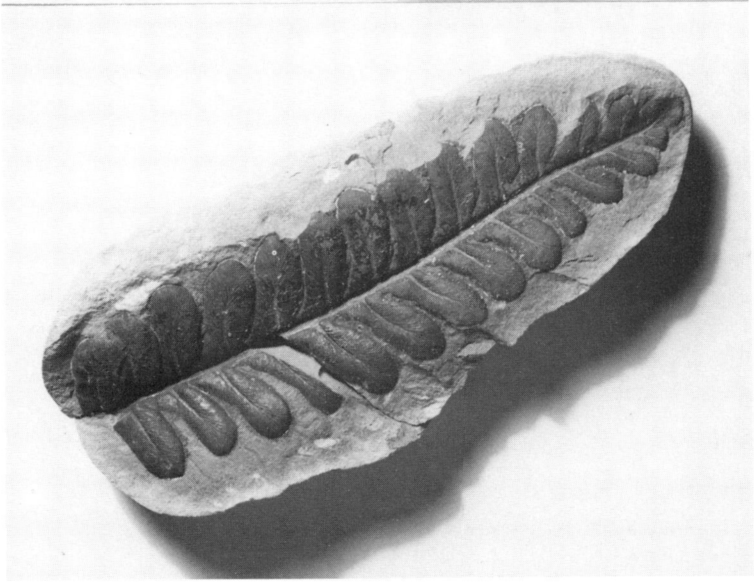

FIGURE 133. (Above)

Alethopteris ambigua Lesquereux. (X 1)
ISM 14242

FIGURE 134. (Below)

Alethopteris sullivanti Lesquereux. (X 3/5)
ISM 14259

once, and a few twice. This species was formerly included in the genus *Callipteridium* Weiss.

Desmopteris longifolia Presl (Fig. 135) has pinnules which are very long and linear and are not decurrent along the rachis. They vary in length from one and a half to three and a half inches, and are from a quarter to three-eighths of an inch wide. The veinlets usually fork once. Sometimes minute teeth may be observed along the margins, corresponding with the ends of the veinlets.

Desmopteris integra Gothan (Fig. 136) differs from *D. longifolia* in having an undulating border, and more complex venation. Some veinlets in each pinnule fork once only, while others fork from two to four times.

FIGURE 135

Desmopteris longifolia Presl. (X ¾)

ISM 14355

Neuropterid Group

Neuropteris scheuchzeri Hoffmann (Figs. 8 & 137) is characterized by a trifoliate pinna consisting of one large, lanceolate, median pinnule, and two smaller, rounded, lateral pinnules. These frequently are found disconnected, however. The larger pinnules, which vary from one and a half to six inches in length, by one-half to one and a half inches in width, may be sharply pointed or obtusely tipped. The small, lateral pinnules are nearly round and seldom measure more than a half inch in diameter. The pinnule bases are cordate, or heart-shaped. The veinlets originate from a midrib, and arch widely toward the margins, forking from one to five times. Numerous short, heavy hairs usually are present on the surface. They may be mistaken for veins or scratches at first glance, but further observation reveals that they generally lie at an oblique angle to the venation.

Neuropteris clarksoni Lesquereux (Fig. 138) has linear-lanceolate pinnules which are usually somewhat longer in relation to their width, and more pointed, than those of *N. scheuchzeri.* The pinnae do not have the trifoliate character of *N. scheuchzeri,* nor do they have hairs on the surfaces of the pinnules. The venation is similar, the veinlets forking from one to five times.

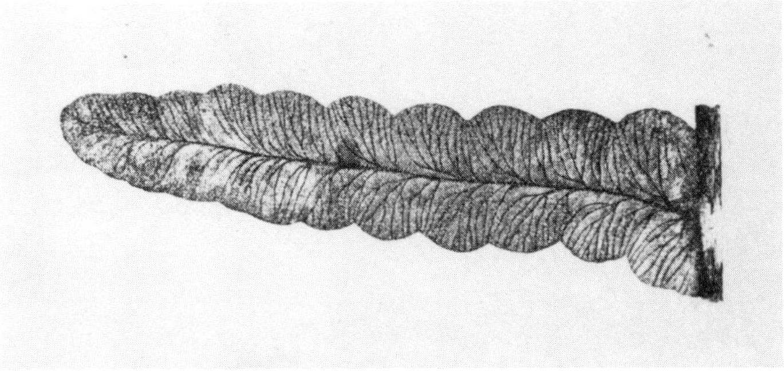

FIGURE 136

Desmopteris integra Gothan. (X 1½) Collected in the Saar, Germany (From Potonié).

Neuropteris flexuosa Sternberg (Fig. 139) has a distinctly flexed, or bow-shaped, leaf varying in length from a half to one and a half inches. The veins are heavier and farther apart than in the preceding species, and fork from one to five times. The terminal pinnule of a pinna is usually much larger than the lateral pinnules.

FIGURE 137

Neuropteris scheuchzeri Hoffmann. A complete pinna showing the large terminal pinnule and the two small lateral leaflets at the base. (X 7/10)
ISM 14449

Neuropteris plicata Sternberg (Fig. 140) is very closely allied to *N. flexuosa,* and some authors unite the two species. In contrast, the pinnules of *N. plicata* seem to be characteristically shorter in relation to their width, and the lower portion of the base is drawn out into a rounded auricle on all but the smaller pinnules.

Photograph by George Langford

FIGURE 138 ISM 14779

Neuropteris clarksoni Lesquereux. (X ½)

FIGURE 139

Neuropteris flexuosa Sternberg. (X 1)

ISM 14454

Neuropteris tenuifolia Schlotheim (Fig. 141) has many of the same characteristics as *N. flexuosa.* It is longer in relation to its width, however, and does not always have the characteristic flexed shape of the pinnules. The venation is similar, but the terminal pinnule is usually smaller than the lateral ones.

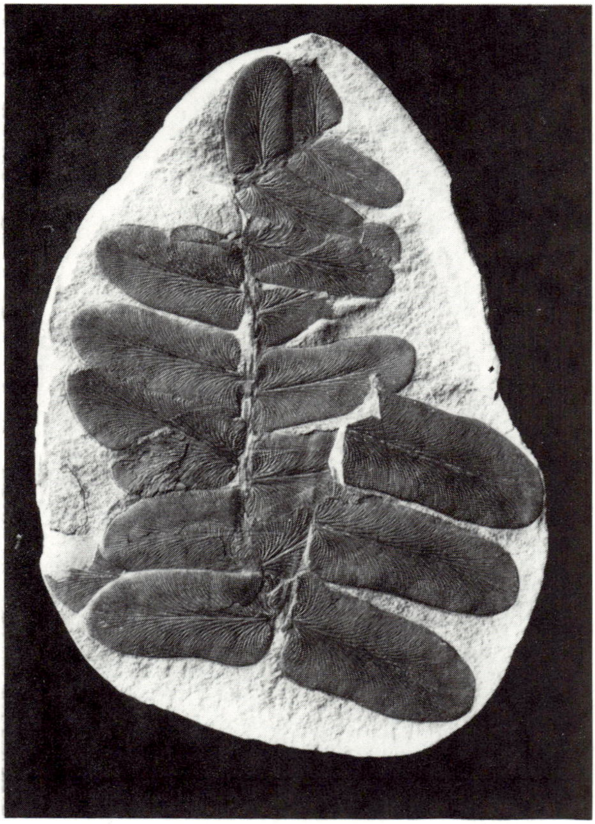

FIGURE 140

Neuropteris plicata Sternberg. (X 1)
ISM 14468

Photograph by George Langford

Neuropteris gigantea Sternberg (Figs. 142 & 143), in spite of its name, has medium sized pinnules oblong in shape with nearly parallel sides and obtusely rounded apexes. They are usually spaced quite closely together and nearly at right angles to the rachis so as to cause an overlapping upon one another after the manner of shingles or tile roofing. The venation is similar to that of *N. flexuosa* and *N. tenuifolia.* The relative size of the terminal pinnule is in marked contrast to those of the two latter species, however, as it is generally much smaller than the lateral pinnules of the same pinna.

Photograph by George Langford

FIGURE 141
Neuropteris tenuifolia Schlotheim. (X 1)
ISM 14483

Neuropteris rarinervis Bunbury (Fig. 144), as its name indicates, has few and widely spaced veinlets as compared to the other species of *Neuropteris.* The veinlets, which fork two to three times, are quite thick, appearing almost double in strength. The pinnules are oblong in form and are quite small, averaging from a quarter to a half inch in length. Near the bases of many of the fronds in this species, there were large orbicular leaves with radiating, but widely spaced, venation. When found detached, it is often advisable to refer such leaves to the genus *Cyclopteris.*

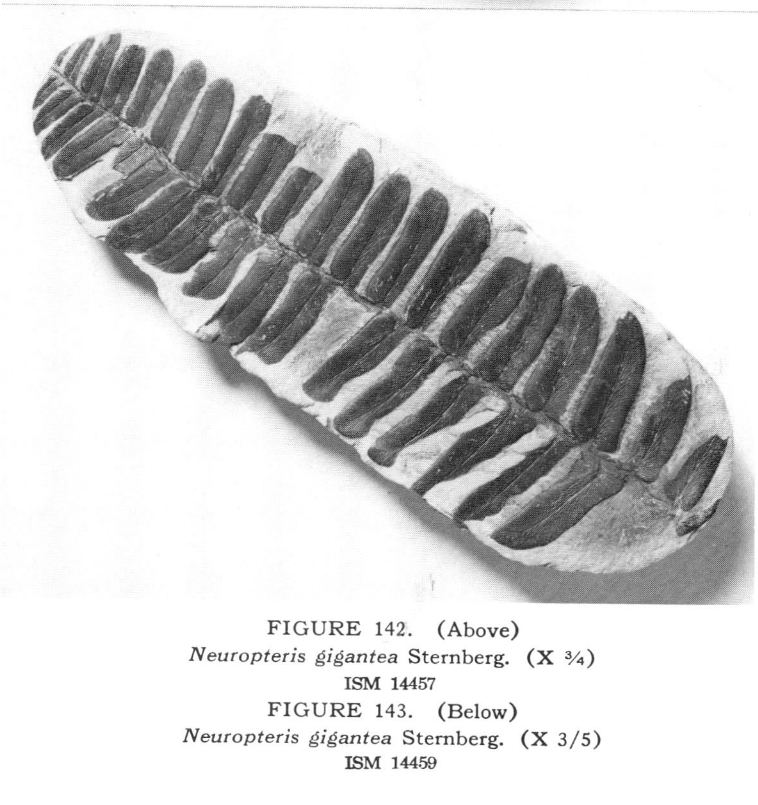

FIGURE 142. (Above)
Neuropteris gigantea Sternberg. (X 3/4)
ISM 14457
FIGURE 143. (Below)
Neuropteris gigantea Sternberg. (X 3/5)
ISM 14459

FIGURE 144
Neuropteris rarinervis Bunbury. (X ½)
ISM 14475

FIGURE 145

Neuropteris heterophylla Brongniart. Tip of a frond, showing enormously enlarged terminal pinnules on each pinna. From Murphysboro, Illinois. (X ½)
ISM 1624

Neuropteris heterophylla Brongniart (Figs. 145 & 146) has pinnules which do not exceed three-quarters of an inch in length, and have a more rounded appearance than those of *N. rarinervis.* The veinlets fork two or three times, but are much finer and more closely spaced than in the latter species.

FIGURE 146

Neuropteris heterophylla Brongniart. (X 3/5)
ISM 14460

Neuropteris fimbriata Lesquereux (Fig. 147) has large pinnules with delicately fringed upper margins. The veinlets fork two to five times, one nerve extending to the tip of each fringe. There is some question as to whether this is a distinct species, or represents merely a lacerated condition of the pinnules due to partial decay before becoming fossilized. Although not abundant, this form has been found throughout much of North America and Europe, and until definitely proved otherwise, it is listed as a distinct species.

Cyclopteris trichomanoides Sternberg (Figs. 148 & 149) includes numerous types of cyclopteroid, orbicular, or reniform leaves of the Neuropterid group which occur isolated from their parent plants and cannot be assigned to other definite species. The veins radiate from the point of attachment, fork several times, and arch to the margins. Nearly every leaf is slightly different from the next, and various species have been made by authors to accommodate such differences. Such species names are of little value, and it is

FIGURE 147

Neuropteris fimbriata Lesquereux. A portion of a pinnule showing the characteristic fringed edge. (X 1) Collected near Galesburg, Illinois.

Photograph by author

more satisfactory to refer all detached specimens to *C. trichomanoides* which has priority as a form name.

Mixoneura ovata Hoffmann (Fig. 150) has a midrib composed of several strands which enter at a single point and give rise to veinlets which fork only once or twice. The pinnules are relatively small. They are slightly contracted at the base, giving them a distinctly ovate shape; hence their name.

Mixoneura jenneyi White (Fig. 151) is a very distinctive form. It has pinnules which are attached by the lower halves of their bases. Numerous veins enter the pinnule from the rachis. In addition to a midrib composed of several strands,

FIGURE 148

Cyclopteris trichomanoides Sternberg. A peculiarly shaped basal pinnule from the frond of a *Neuropteris*. (X 1)
ISM 14350

FIGURE 149

An orbicular form of *Cyclopteris trichomanoides* Sternberg. (X 1)
ISM 14334

FIGURE 150

Mixoneura ovata Hoffmann. (X ¾)
ISM 14443

FIGURE 151

Mixoneura jenneyi White. (X ½)
ISM 14462

Photographs by George Langford

FIGURE 152. (Above)

A single pinnule of *Linopteris neuropteroides* Potonié showing the net-like venation. (X 1)

ISM 14780

FIGURE 153. (Below)

Linopteris muensteri Potonié has the network of the venation much coarser at the middle of the pinnule than at the borders. (X 1)

ISM 14781

several other veins enter along the attached half of the base. These veins leave the rachis at a very narrow angle and nearly parallel the rachis for a short distance before arching abruptly toward the margins. In approaching the borders, the veins divide from one to five times. The lateral pinnules are tongue-shaped, and the terminal one is larger than the others.

Linopteris neuropteroides Potonié (Fig. 152) has oblong pinnules comparable in shape to those of *N. gigantea.* However, its venation consists of a network of anastomosing nerves, and in this respect is distinct from all forms of *Neuropteris* and *Mixoneura.* The meshes of the network are about the same size over the entire pinnule.

Linopteris muensteri Potonié (Fig. 153) may be distinguished from *L. neuropteroides* by its more linear and pointed pinnules. Likewise, the meshes of the venation are much coarser near the mid-portion of the pinnules.

Odontopterid Group

Odontopteris wortheni Lesquereux (Fig. 154) has pinnules which are more or less tongue-shaped, are slightly constricted near their bases, and are attached by the entire breadth. The lower pair of pinnules on each pinna are enormously enlarged. There is no midrib; several veins enter the pinnule from the rachis, fork several times, and arch toward the margins. Numerous small, adpressed hairs may usually be seen on the surface.

Odontopteris subcuneata Bunbury (Fig. 155) differs from *O. wortheni* in having the lower base of most pinnules drawn out into a half-round auricle. Hairs do not occur on the pinnules of this species.

Odontopteris aequalis Lesquereux (Fig. 156) has oval pinnules which are widest near the middle. Sometimes they are placed along the rachis at a slight distance from each other so that their sides do not touch. There is no midrib. The veins are few, are nearly parallel at the base, and diverge slightly upwards as they approach the margins. They usually fork once above the middle.

FIGURE 154

Odontopteris wortheni Lesquereux. (X 1)

ISM 14495

FIGURE 156

Odontopteris aequalis Lesquereux. (X 4/5)

ISM 14597

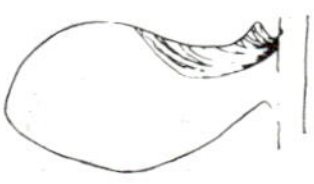

FIGURE 155

Diagram of *Odontopteris subcuneata* Bunbury, showing characteristic half-round auricle at base of pinnule. (X 2)

CHAPTER VII

Gymnospermous Seeds

IMPRESSIONS of many kinds of seeds are found in the fossiliferous shales. Some resemble nuts, and others look like the winged seeds of various modern trees. They usually occur detached, and belonged to either the *Cordaites* trees or to seed-ferns. Seeds attached to identifiable stems or leaves are so rare that the finding of such a specimen is usually of considerable scientific importance. Such specimens found by laymen should be immediately turned over to a museum or university for detailed study.

As in the case of leaves and stems, detached impressions of seeds must also necessarily be classified on the basis of external form. Such a situation is unfortunate from a biological standpoint as many forms which appear identical externally may be quite different structurally. Consequently, it is usually impossible to assign impressions of seeds to genera and species based on structural characteristics. Generic "form" names are therefore being used in this handbook, but the "structure" names will also be indicated when they fall within the respective form genera.

Pachytesta Brongniart (Figs. 157 & 161, i) is an unusually large, oval seed, three to five inches long. It is radially symmetrical, with numerous ribs and grooves extending along its entire length. One end is slightly flattened where it was formerly attached to the stem, and the other end terminates in a blunt point.

Holcospermum Nathorst (Figs. 158 to 160 & 161, f) now includes many seeds which were formerly classified under the genus *Rhabdocarpus* (see below). They are radially symmetrical, have numerous longitudinal ribs and grooves such as those of *Pachytesta,* but are much smaller, averaging from one-half to two inches in length. Small seeds of this type have been found imbedded close to leaflets of *Neuropteris*

FIGURE 157

Pachytesta Brongniart.
(X 1)
ISM 14498

scheuchzeri, and it is possible that one form of *Holcospermum,* at least, may have belonged to this seed-fern.

Rhabdocarpus Berger (Fig. 161, h) is a globular, or flask-shaped seed having a pointed extension at its apex, suggesting a green onion in shape. It is about an inch or inch and a half long and has numerous longitudinal ribs and grooves. As now classified, this genus includes a single species, *R. tunica-*

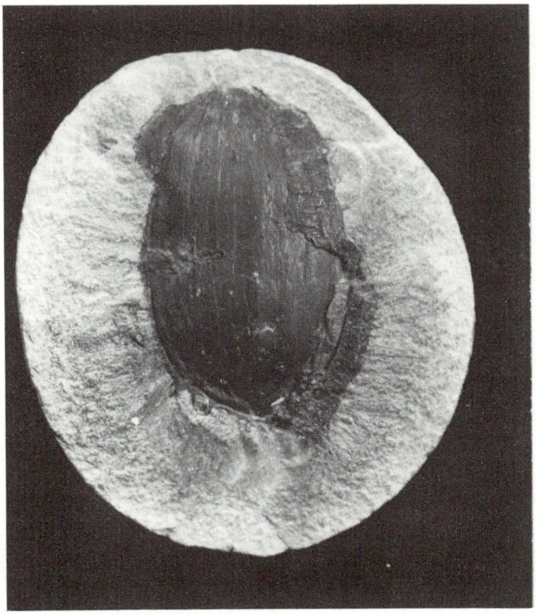

FIGURE 158

Holcospermum Nathorst, showing external suface of the husk. (X 1)
ISM 14782

Photograph by George Langford

FIGURE 159

Holcospermum Nathorst, showing surface of seed minus the husk. (X 1)
ISM 14783

Photograph by George Langford

tus Berger, which has been described on the basis of internal structure.

Codonospermum Brongniart (Fig. 161, j) is an ellipsoidal seed which is differentiated into an upper and lower region of

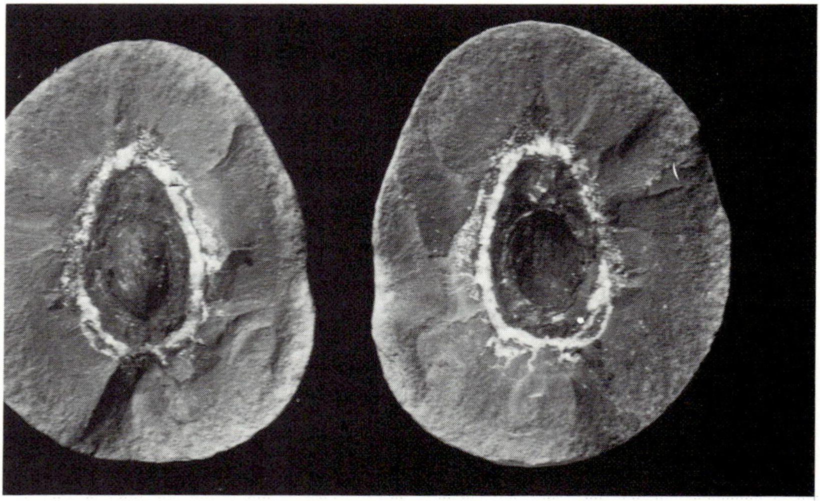

Photograph by George Langford

FIGURE 160

Holcospermum Nathorst, showing seed surrounded by husk. (X 1)

ISM 14784

differing diameters. It usually has eight ribs extending along its entire length. These seeds average about an inch in length.

Lagenospermum Nathorst (Fig. 161, e) includes very small oval or elliptical seeds which are not more than a

---

FIGURE 161

Gymnospermous Seeds.

a. *Trigonocarpus* Brongniart (X ¾);
b, c. *Samaropsis* Goeppert, widened and elongated forms (X 1);
d. *Cordaicarpus* Geinitz (X 1);
e. *Lagenospermum* Nathorst (X 2);
f. *Holcospermum* Nathorst (X ¾);
g. *Codonotheca* Sellards (X 1);
h. *Rhabdocarpus* Berger (X 1);
i. *Pachytesta* Brongniart (X 1);
j. *Codonospermum* Brongniart (X 1).

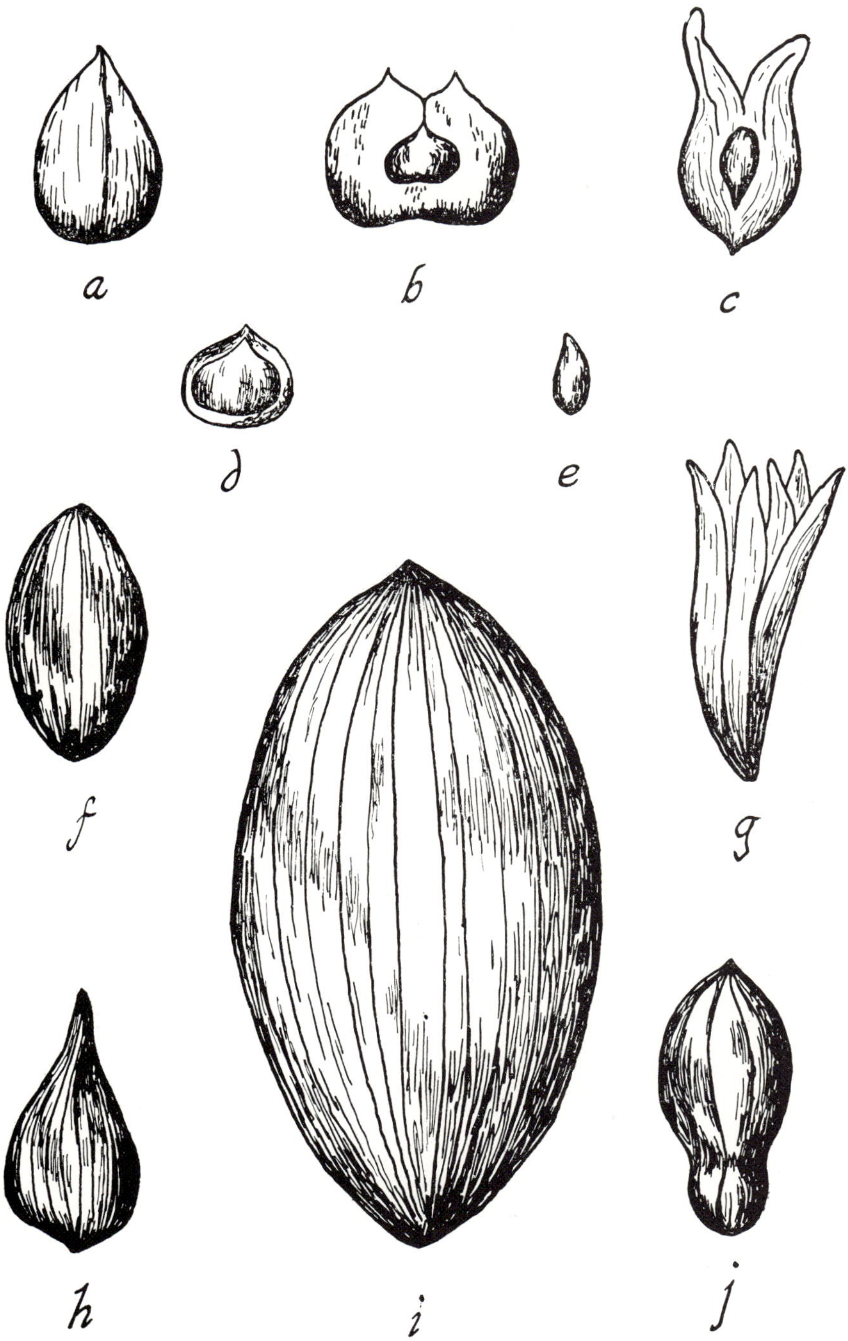
a
b
c
d
e
f
g
h
i
j

quarter of an inch in length. Some are smooth and others have numerous ribs and grooves along their entire or partial length. On the basis of internal structure, Williamson has placed some forms in the special genera *Lagenostoma,*

FIGURE 162

Trigonocarpus Brongniart. Found near Grayville, Edwards County, Illinois. (X 2)

ISM 1781

FIGURE 163

Samaropsis Goeppert. (X 1)

ISM 14533

Physostoma, and *Conostoma.* Of course, only petrified seeds which are capable of being sectioned and studied internally can be assigned to these latter "structure" genera.

Trigonocarpus Brongniart (Figs. 161, a & 162) includes medium sized seeds ranging from three-quarters of an inch to

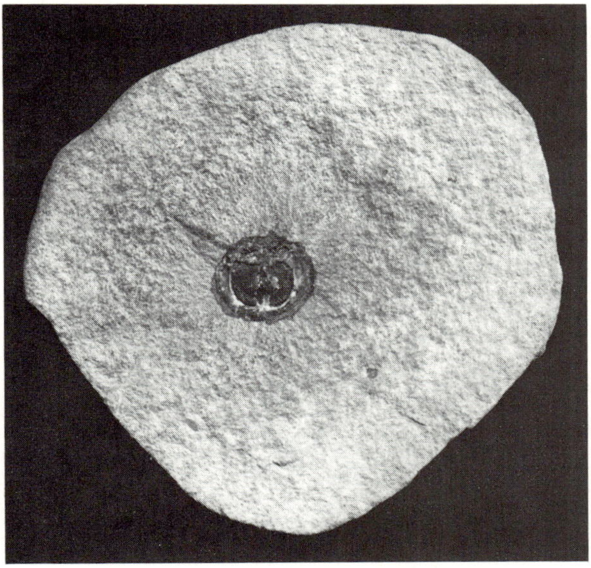

FIGURE 164

Cordaicarpus Geinitz has the seed rounded by a narrow wing. (X 1)

ISM 14785

Photograph by George Langford

FIGURE 165

Codonotheca Sellards. (X 2)

Courtesy Illinois State Geological Survey.

two inches in length. As the name indicates, the outer shell is divided longitudinally into three equal valves, and these are marked by prominent ridges at the suture lines. The intermediate surfaces are usually smooth, but may sometimes be marked by distinct longitudinal striations. Seeds of *Trigonocarpus* have been found close to the leaves of *Alethopteris lonchitica,* and it is possible that some forms may belong to this plant. The genus includes the special genus *Tripterospermum* Brongniart, which is differentiated on the basis of internal structure.

Samaropsis Goeppert (Figs. 161, b, c & 163) includes impressions of flat, more or less circular seeds. The seed is surrounded by an oval or heart-shaped border, or wing, called the sarcotesta. As in the case of some modern seeds, this wing probably aided the dispersal of the seeds by the wind. *Samaropsis* seeds were borne by some of the *Cordaites* trees, and there is evidence that certain forms belonged to some of the seed-ferns.

Cordaicarpus Geinitz (Figs. 161, d & 164) includes seeds similar to *Samaropsis* but which do not possess a definite sarcotesta, or show only a very narrow one. These also belonged to *Cordaites* trees or to seed-ferns. The genus includes the special genus *Cardiocarpus* Brongniart, which is differentiated on the basis of internal structure.

Codonotheca Sellards (Figs. 161, g & 165) has the form of a stalked cup consisting in the basal portion of a stout axis, and in the upper portion of six linear segments. These are united basally to form the sloping surface of the cup. They average an inch to an inch and a half in length.

Carpolithus Linnaeus (sometimes spelled *Carpolithes*) includes casts and impressions of large and small detached seeds which lack regular ribs, grooves, wings, or other distinguishing features, and consequently cannot be assigned to any of the above more clearly defined genera. Some belonged to seed-ferns, and others to the *Cordaites* trees.

CHAPTER VIII

THE BIRTH OF PALEOBOTANY

WHEN WE SURVEY all that is known concerning the geological history of our earth, it is difficult to realize that comparatively little was known on this subject more than a few generations ago. Geology as a science had not yet come into existence. The subdivision, paleobotany, must be considered as an even younger science than general geology. Although the ancients were familiar with many kinds of fossil shells and other invertebrate animal remains, it seems that they were practically unacquainted with any form of vegetable fossils.

This lack of knowledge concerning fossil plants seems to have continued throughout the middle ages down to the thirteenth century. Because most humans are naturally curious, it is difficult to believe that such common objects as petrified wood or leaf impressions should not have been observed during the intervening ages. We prefer to believe, therefore, that such remains were recognized from time to time, but that no records were made of such observations, or if so, that they have since become lost.

Agricola[1] has been credited with being the first to mention the existence of petrified wood in his writings of 1558. However, a passage occurs in Magnus[2] which indicates that this subject had attracted attention as early as the thirteenth century, after which it apparently received no further consideration until Agricola mentioned it in his writings. These were followed by accounts of Matthiolus in 1564, and by

---

[1] Georgius Bauer Agricola, De Natura fossilium, Lib. VII, pp. 324, 328. 1558.

[2] Beati Alberti Magni, De Mineralibus, Tractatus I, Caput VII, Opera, Tome II, p. 216. Lugduni, 1651.

Gesner in 1565, in which they described specimens of petrified wood which came into their possession.[1]

The first account of leaf impressions was written by Johannes Kentmann[2] in 1565, at which time he described impressions that had been formed by incrustations of mineral matter in warm springs deposits. The earliest account of plant leaves in an actual rock formation was made in 1664 by Johann Major.[3] This was followed in 1699 by the work of Edward Llhwyd[4] in which he not only described, but also pictured, a number of ferns from the coal age rocks of England.

Naturally the recognition of organic remains in rock formations brought forth speculations as to their origin and mode of occurrence. Vague and unreal notions prevailed up to the eighteenth century wherein the phenomena were attributed to various causes. Some writers believed that these plant remains were the work of stone-making spirits. Others held they were cast-off creations of God. Still others advanced the theory that such fossils were formed by the action of petrifying juices in the earth's crust upon living plants. Some refused to postulate any theories, holding that they were merely "freaks of nature." A bit later these explanations were supplanted almost entirely by the "Flood Theory," in which it was held that all fossils consisted of the remains of life on earth prior to the occurrence of Noah's flood as recorded in the Bible. It was thought that the dead remains were washed about in the flood of waters, and then were left stranded on or near the land surfaces upon the retreat of the waters. Martin Luther[5] seems to be the first to have sug-

[1] Lester F. Ward, Sketch of Paleobotany, United States Geological Survey Annual Report No. 5, p. 388. Washington, 1885.

[2] J. Kentmann, Nomenclatura rerum fossilium, etc., Lib. VI, fol. 38. Lapides & Tiguri, 1565.

[3] Johann David Major, Lithologia curiosa, sive de animalbus et plantis in lapides versis. Jena, 1664.

[4] Eduardi Lluidii, Lithophylacii britannici ichnographia. London & Leipzig, 1699.

[5] Martin Luther, Gründliche and Erbauliche Auslegung des Ersten Buchs Mosis, Band I, col. 176. Halle, 1739.

gested that proof of the great deluge could be seen in the form of wood hardened into stone in the vicinity of certain mines and smelting mills. In 1696, a volume was published by John Woodward[1] in which he held that all the solid parts of the earth's crust were loosened by the flood and were mingled by the surging waters. As the flood waters receded, everything resettled to form a new land surface according to the specific gravity of the various pieces, thus accounting for the finding of plant and animal remains in the rock strata.

The greatest expounder of the flood theory was Johann Scheuchzer, whose work appeared in 1709 under the title, "Herbarium Diluvianum." It was in his honor that Hoffmann later named the fossil plant species, *Neuropteris scheuchzeri.* Scheuchzer did not accept Woodward's entire hypothesis, but believed that the plants and animals were buried in the mud and slime resulting from the deluge and that they later underwent the process of petrifaction while buried. The diluvian theory rapidly became the prevailing doctrine and most of the scientists of the eighteenth century accepted it, or at least did not openly oppose it. Scheuchzer considered the fossils as ordinary organisms still to be found living on earth, and he named them in accordance with those described in standard biological volumes of his time.

Very soon, however, students began to compare the best preserved fossil specimens with living plants as evidenced by a memoir on coal plants published in 1718 by Antoine de Jussieu.[2] In this treatise he pointed out the resemblance of his specimens to the leaves of living tropical plants. As these studies continued, it became apparent to workers that many plant fossils were vastly different from any forms known to be growing anywhere on the earth today. It was Blumenbach[3] who finally announced that the fossils were not only

[1] John Woodward, An Essay Towards a Natural History of the Earth and Terrestrial Bodies. London, 1695.

[2] A. de Jussieu, Examen des causes des impressions des plantes marquées sur certaines pierres des environs de Saint Chaumont. Paris, 1718.

[3] J. F. Blumenbach, Beiträge zur Naturgeschichte. 1790.

antediluvial, but that there had been a great succession of faunas and floras inhabiting the earth long before the advent of man. The inadequacy of the flood theory in explaining the observed facts inspired further research, thereby inaugurating truly scientific investigations which have continued to the present day.

Paleobotany, as a science, dates only from the beginning of the nineteenth century. Nevertheless, as Ward[1] has so aptly stated, its beginning was preceded by a long, fading twilight which extended entirely through the eighteenth century.

The foundations of scientific paleobotany were laid in Germany in 1801 by Ernst Fredrich, Baron von Schlotheim when he published in Hoff's Magazine[2] his "Abhandlung über de Kräuter-Abdrücke in Schieferthon und Sandstein der Steinkohlen-Formation." This was followed in 1804 and 1820 by volumes in which he accurately described and pictured many Carboniferous plants. These works of Schlotheim are the earliest strictly scientific records known in paleobotany.

Next in order appears the name of Kasper Maria, Graf von Sternberg. Although living contemporaneously with Schlotheim, Sternberg's first contribution did not appear until three years after Schlotheim's first, and his most important work was not finished until after the last of Schlotheim's works were published. He was president of the Bohemian National Museum, and his chief interest was in botany, but many of his publications concerned fossil plants. His important work appeared in several volumes between the years 1820 and 1838. This work was translated into French by the Compte de Bray under the title, "Essai d'un exposé géognostico-botanique de la flore du monde primitif." It is accompanied by many plates of hand-painted water colors which are so well done that the species can be readily recognized for comparison with fossils even today.

---

[1] L. F. Ward, op. cit., p. 368.
[2] Hoff's Magazine, Vol. 1, pp. 76-95. Leipzig, 1801.

Although Schlotheim and Sternberg are regarded as pioneers in the science, Adolphe Théodore Brongniart is universally considered to have been the true founder of paleobotany. His first memoir was published in 1822 by the Museum of Natural History of Paris,[1] and was a work of considerable importance. The year 1828 is considered most eventful, for in it was issued Brongniart's famous "Prodrome d'une Histoire des végétaux fossiles," and later in the same year he began another well-known book entitled merely "Histoire des végétaux fossiles." Numerous other volumes and memoirs appeared until 1849. Throughout his work, Brongniart maintained that fossils of plants were, none the less, true plants, and that as fast as they became known they should be assigned to their proper places in the plant series. The systematic manner in which he organized his scientific material made him the highest authority of his time on fossil plants. His "Histoire," with its more than a hundred and ninety well executed plates, was reprinted in 1915 by W. Junk of Berlin, and is one of the most useful of the early paleobotanical publications.

The first important work to appear in the English language was done by John Lindley and William Hutton, entitled "The Fossil Flora of Great Britain." It appeared in quarterly numbers between the years 1831 and 1837.

The most prolific writer in the field of paleobotany was Heinrich Robert Goeppert. He produced more than a hundred writings over a period of half a century from 1834 to 1884.

Some of the most important work ever accomplished in paleobotany was done by Wilhelm Philip Schimper. His "Traité de paléontologie végétale" was issued in a series of three volumes and an atlas of plates during the years 1869 to 1874. It was, without doubt, the greatest contribution to the science of paleobotany which had appeared up to that time.

---

[1] A. T. Brongniart, Sur la classification et la distribution des végétaux fossiles en général. Paris, 1822.

It not only included a complete compilation of the previous work of other writers, but contained a great amount of new material as well. Its organization was highly scientific and it was, in fact, a complete manual of paleobotany.

The earliest known plant fossils from America were obtained by Ebenezer Granger[1] in 1821 from the coal fields near Zanesville, Ohio. He pictured thirteen plant forms which he compared to those previously known in Europe, but he did not name or describe his specimens. He did, however, call attention to the fact that they appeared tropical in nature.

The first work of consequence to be accomplished in North American paleobotany was done by Sir William Dawson, a native of Nova Scotia. His publications appeared for many years, beginning in 1845. His "Geological History of Plants," published in 1880, was the first American textbook in paleobotany.

At about the same time, paleobotanical studies in the United States were begun by Leo Lesquereux, a native of Switzerland. The political upheavals in Europe during 1847-48 forced many scientists to abandon their native lands, several emigrating to America. Among them was Lesquereux. In 1845, previous to his departure, the canton of Neuchâtel, Switzerland was interested in the protection of peat bogs for the purpose of supplying fuel for the poorer classes. In this connection the government offered a prize for the best paper on the origin and preservation of peat. Lesquereux completed and won the prize. This led him into the study of fossil plants.

After his arrival in the United States he became associated as paleobotanist with numerous state and territorial geological surveys which were then being organized from time to time. His work for surveys in the Lake Superior region, Ohio, Arkansas, Pennsylvania, Illinois, Kentucky, and elsewhere gave him a vast knowledge of the Carboniferous plants

---

[1] Ebenezer Granger, Notice of vegetable impressions on the rocks connected with the coal formation of Zanesville, Ohio. American Journal of Science & Arts, New Haven, Vol. III, 1821, p. 5, plates I, II.

of America. The results of his work are included in the Government reports of the various surveys of that time. Most important of these is a series of volumes dealing with the coal flora of Pennsylvania, the first of which appeared in 1858. This was followed at the time of the second geological survey of the state in 1879-80 by two additional volumes and an atlas of plates. In the meantime, Lesquereux had been employed with the first geological survey of Illinois, and Volumes II and IV of the report, issued in 1866 and 1870, respectively, are largely devoted to his work on the plant fossils of Illinois. Many of the plant species which he found in Illinois were new to science, and formed a nucleus for the present paleobotanical collection of the Illinois State Museum, which was first organized to house the geological material collected by the original state survey.

This brief sketch of the birth of paleobotany in Europe and America necessarily omits the names of many eminent scientists in that field. Only the foremost of the pioneers could be mentioned here. Ever since its earliest beginnings, paleobotany has attracted many workers into its fields of research, and to include even a reasonable number of the total would require very many pages. At the present time, the workers in paleobotany are comparatively few. In 1924, Wieland[1] numbered those in definitely recognized paleobotanic positions throughout the world at about a score. Since that time there has been a slight increase; however, the addition of new workers in the field has been offset almost equally by deaths among the older ones.

[1] G. R. Wieland, Recent Achievement in Paleobotany, Science, September 12, 1924, p. 233.

CHAPTER IX

Interpreting the Past

Those of us who are interested in investigating the past history of our earth are confronted with a kaleidoscope of ever-changing scenes. Our first picture shows the earth as a whirling ball, flying through space like a gigantic spinning top. At times its surface is torn by the eruption of belching volcanoes, and across the midnight heavens is reflected the fiery glow from unplumbed depths of red-hot brimstone. At other times we see the globe churned by the rise and fall of continents, or its surfaces swept by the waves of long-lost seas. Time and again we see portions of the lands covered by perpetual ice and snow. Sometimes we see the earth rocked by tremendous earthquakes, and then for long ages everything is serene and quiet. Throughout all this long, eventful history, we witness the appearance of life upon earth, and follow its gradual culmination to present-day forms. In our survey of by-gone ages, we discover evidences of a shifting balance of power among different forms of life, and our feeble minds grope for adequate explanations of these ever-changing events.

The story told by plant fossils adds an important bit to our knowledge of past events. Plants now insignificant and few in number, such as the ground-pie, *Lycopodium,* are found to be descendants of a long line of ancestors extending back into remote ages when they formed the dominant growth of primeval forests. Others, like the seed-ferns, have left no descendants whatever. Still others, the ferns themselves, were able to hold their own through long ages and have come down to us in an unbroken line. We can only speculate vaguely, as yet, on the causes of success or failure which have determined such events.

As we have seen, the flora of the great coal age was composed of trees and plants which more approximately compare with those of our present tropical regions than with any other. The trees of the Carboniferous Period do not exhibit annual growth rings in their stems, and this too is an indication of more or less uniform growing conditions without seasonal fluctuations. This situation excites our interest because the great coal deposits of the Pennsylvanian Periods are situated in the northern hemisphere and extend from localities within the Arctic Circle down into the temperate zone. Vast coal deposits, with their associated fossils of tropical types of plants, are found in many of the Arctic islands, at Spitzbergen, in Greenland, Siberia and Alaska. From this evidence it would seem that a warm, uniform climate must then have existed over the greater portion of the earth's surface.

Although very difficult to imagine, it is possible that such a climate could have existed during the coal age and would account for the finding of plant remains in the Arctics. However, the problem is not so easily answered. In addition to a favorable climate, plants require an abundance of sunlight for luxuriant growth. Our explanation has failed to show how the coal-forming plants could have received sufficient sunshine to have grown continuously without showing even an indication of seasonal growth rings in latitudes where the polar nights are several months long.

In attempting to explain this riddle, a theory suggesting a possible change in the location of continental land masses has been applied. This hypothesis, called the Continental Drift Theory, was proposed by the German scientist, Alfred Wegener, about 1920. His theory suggests that the continents were once a part of a single larger continent that existed in more southern latitudes where luxuriant plant life was possible. This great continent is represented as having floated upon a plastic sub-stratum. At some time in more recent geologic history, this great continent broke into several sections which drifted apart and formed the continents as we know them today. This theory, which radically opposes the previous conception of the stability of continental masses, has not met with a uniformly favorable reception among geologists. However, nothing else so well explains the wide distribution of the plant deposits of the coal age. The theory is relatively new, and only time and physical and biological evidence will ultimately tell whether or not it has practical merit. The work of the scientist is never finished and he never knows when some bit of evidence, newly discovered or previously considered unimportant, may furnish the clue to a more thorough understanding of the profound changes which have affected our earth and the evolution of life upon it.

BIBLIOGRAPHY

For a more detailed description of the various plant forms and related subjects than can be covered in this handbook, the reader is referred to the following publications which may be obtained at the larger public libraries and museums:

Andrews, Henry N. Jr. Ancient plants and the world they lived in. Comstock Publishing Co., Ithaca, N. Y., 1947.

Arber, E. A. N. The natural history of coal. The University Press, Cambridge, 1911.

Arnold, Chester A. An introduction to paleobotany. McGraw-Hill Book Company, New York, 1947.

Arnold, Chester A. Fossil flora of the Michigan coal basin. Contributions from the Museum of Paleontology, Vol. VII, No. 9, University of Michigan Press, Ann Arbor, 1949.

Berry, E. W. A sketch of the origin and evolution of floras. Annual Report of the Smithsonian Institution for 1918, Washington.

Brongniart, Adolphe T. Histoire des végétaux fossiles. Paris, 1828. Reprinted by W. Junk, Berlin, 1915.

Crookall, R. Coal measure plants. Edward Arnold and Company, London, 1929.

Dahlgren, B. E. A forest of the coal age. Geology Leaflet No. 14, Chicago Natural History Museum, Chicago, 1933.

DuToit, A. L. Our wandering continents. Oliver & Boyd, London, 1937.

Hirmer, Max. Handbuch der Paläobotanik. R. Oldenbourg, München and Berlin, 1927.

Jackson, B. D. A glossary of botanic terms. J. B. Lippincott Company, Philadelphia, Fourth Edition, 1928.

Janssen, Raymond E. Some fossil plant types of Illinois. Scientific Papers Vol. I, Illinois State Museum, Springfield, 1940.

Jongmans, W. J. Anleitung zur Bestimmung der Karbonflanzen West-Europas. Band I. Memoirs of the Government Institute for the Geological Exploration of the Netherlands, 's-Gravenhage, 1911.

Jongmans, W. J. and Kidston, R. A monograph of the Calamites of western Europe. Text and atlas. Memoirs of the Government Institute for the Geological Exploration of the Netherlands, 's-Gravenhage, 1917.

Kidston, Robert. Fossil plants of the Carboniferous rocks of Great Britain. Memoirs of the Geological Survey of Great Britain, Paleontology, Vol. II, Parts 1 to 6, London, 1923 and 1925.

Knowlton, F. H. Plants of the past. University Press, Princeton, 1927.

Lesquereux, Leo. An enumeration of the fossil plants found in the coal measures of Illinois, with descriptions of new species. Illinois State Geological Survey, Vol. II, Springfield, 1866.

Lesquereux, Leo. Report on the fossil plants of Illinois. Illinois State Geological Survey, Vol. IV, Springfield, 1870.

Lesquereux, Leo. Coal flora of Pennsylvania and of the Carboniferous formation throughout the United States. Second Geological Survey of Pennsylvania, Report P. Three volumes and atlas. Harrisburg, 1879-84.

Noé, A. C. Pennsylvania flora of northern Illinois. Illinois State Geological Survey, Bulletin No. 52, Urbana, 1925.

Potonié, H. and Gothan, W. Lehrbuch der Paläobotanik. Gebrüder Borntraeger, Berlin, 1921.

Renier, Armand. Documents pour l'étude de la paléontologie du terrain houiller. Liége, 1910.

Scott, D. H. Studies in fossil botany. Two volumes. A. & C. Black, Ltd., London, 1909 and 1923.

Scott, D. H. The evolution of plants. Henry Holt and Company, New York, 1911.

Seward, A. C. Fossil plants. Four volumes. The University Press, Cambridge, 1898-1917.

Seward, A. C. Plant life through the ages. Macmillan and Company, New York, 1931.

Walton, John. An introduction to the study of fossil plants. Adam & Charles Block, London, 1940.

Wegener, Alfred L. The origin of continents and oceans. Translated from the third German edition by J. G. A. Skerl. E. P. Dutton and Company, New York, 1924.

White, David. Flora of the outlying Carboniferous coal basins of southwestern Missouri. United States Geological Survey, Bulletin 98, Washington, 1893.

White, David. Fossil flora of the lower coal measures of Missouri. United States Geological Survey, Monograph 37, Washington, 1899.

INDEX